THE SILICON FOREST

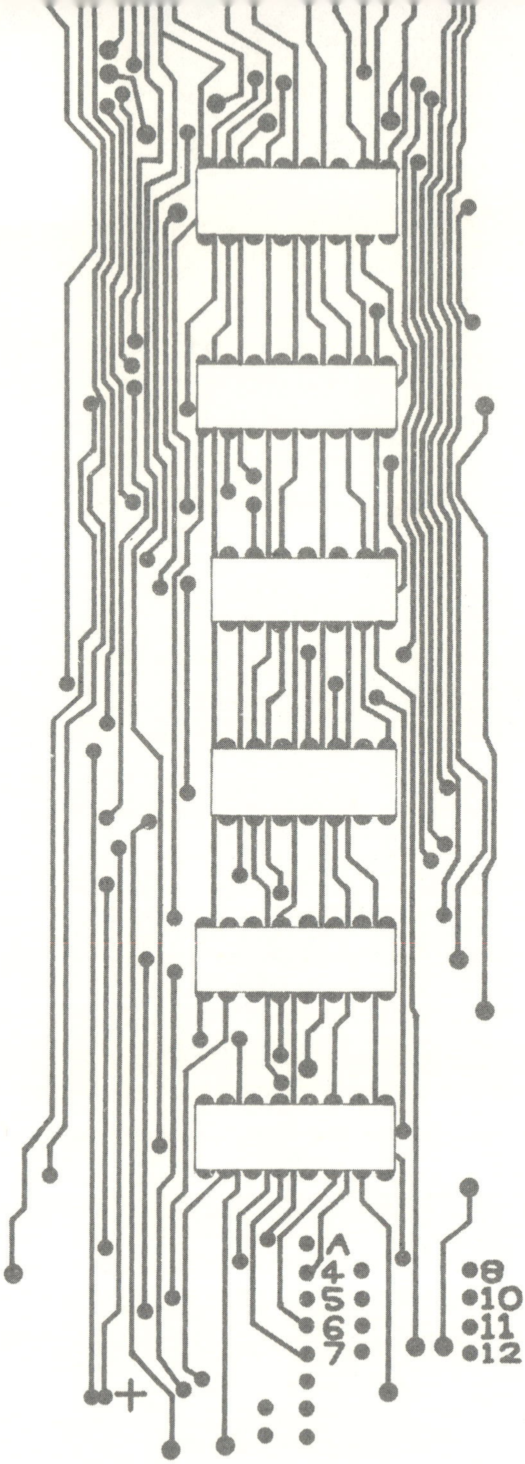
A
4
5
6
7
8
10
11
12
+

THE SILICON FOREST

High Tech in the Portland Area 1945 to 1986

Gordon B. Dodds

Craig E. Wollner

WITH THE ASSISTANCE OF

Marshall M. Lee

Oregon Historical Society Press

Support for this volume was provided in part by funds from the M.J. Murdock Charitable Trust as part of the Jack Murdock Publication Series on the History of Science and Exploration in the Pacific Northwest.

Library of Congress Cataloging-in-Publication Data

Dodds, Gordon B. (Gordon Barlow), 1932–
The silicon forest: high tech in the Portland area, 1945–1986 / Gordon B. Dodds and Craig E. Wollner with the assistance of Marshall M. Lee.
p. cm.
Includes bibliographical references and index.
ISBN 0-87595-226-7
1. High technology industries—Oregon Portland Region—History. I. Wollner, Craig, 1943– . II. Lee, Marshall M. III. Title.
HC108.P87D63 1990
338.4'762'000979549—dc20 90-41708

Oregon Historical Society Press
1230 S.W. Park Avenue, Portland, Oregon 97205

The paper used in this publication is recycled and meets the minimum requirements of American National Standard for Information Sciences—Permanence of Paper for Printed Library Materials, ANSI Z39.48-1984.

Printed in the United States of America.

To Bette and John Dodds

&

To Charles and the memory of Fannie Bloomberg

Contents

Preface, ix

PART ONE: THE COMPANIES

ONE Beginnings, 3

TWO Electro Scientific Industries, 10

THREE Tektronix, 26

FOUR Floating Point Systems, 46

FIVE New Ventures, 56

SIX The Supercomputers, 85

PART TWO: THE LARGER COMMUNITY

SEVEN High Technology and Education, 99

EIGHT Politics, 125

NINE The World of Work, 155

TEN The Balance Sheet, 169

Epilogue, 188

A Note on the Sources, 195

Notes, 196

Index, 219

Preface

THIS BOOK TELLS THE STORY of the high-technology industry of Washington County, Oregon, a thriving and ingenious enterprise that has been of prime importance in the state's recent history and in its economic development.

The story begins in the 1940s in Southeast Portland, where Howard Vollum and Jack Murdock created Tektronix and Douglas Strain formed Electro Scientific Industries (ESI). It continues in Washington County, where Tektronix moved in the 1950s and ESI followed in the early sixties. Both before and after their relocation moves, the two companies represented the scientific and engineering creativity, humanity, rapid growth, and community spirit that was to characterize much of the world's high-technology industry.

In 1970 the firms were joined by Floating Point Systems, founded by C. Norman Winningstad, and, later in the decade and in the early eighties, by a score of other new high-technology ventures. Both old and new companies made major contributions in the areas of measurement and instrumentation, computing, and specialized high technology. In the process the industry opened economic avenues for Oregon and stimulated auxiliary business enterprises: advertising and public relations, law, land development, and venture capital. High technology also influenced the educational system of county and state, and it interacted politically with all levels of government.

In our book we give an overview of all these developments, and we draw conclusions about the nature of the industry and about its positive and negative influences on the larger community. Throughout, we have tried to present a synthesis of the high-technology industry for the general reader. We have made no attempt to be comprehensive and have kept intricate matters of technology and finance to a minimum. The most difficult part of our task has been to grasp and illuminate the significance of a rapidly changing industry, whose story often seems as much current event as history.

The most pleasing part of our work has been meeting and learning from many men and women connected with the industry. They have aided us immeasurably in telling its story. (We have identified them here in terms of their affiliations at the time of writing.) We are particularly grateful to Pat McCormick, American Electronics Association; Linda Rasmussen, Communications Workers of America; Richard P. Reiter, Department of Environmental Quality; Bev Buckner, DeLayne Giardini, Meg Kenagy, Harold Lawson, Douglas Strain, Electro Scientific Industries; David Henderson, Emerald Computers; Masakatsu Tomita, Epson Portland; John Bernard, Terry J. Dols, C. Norman Winningstad, Floating Point Systems; Mike Ragsdale, Grubb and Ellis Property Management; George Drumbor, Intel; Steve Karakas, Karakas, VanSickle, and Ouellette; Ralph Wiser, Kulongoski, Durham, Drummonds, and Colombo; Ed Coughlin, Lattice Semiconductor Corporation; Thomas H. Bruggere, Mentor Graphics; Sam C. Smith, M.J. Murdock Charitable Trust; Timothy R. Weyer, NEC America; Paul Harrington, Northeastern University; Dave Kohlmeier, Northwest Instruments; Denny Miles, Jim Sellers, Oregon Governor's Office; F. Paul Carlson, Margaret Day, Norman Eder, the late Richard Elliott, Bert Gredvig, Frederick Holmes, James Huntzicker, James Hurst, Georgiana Johnsrud, Eileen Lawless, Oregon Graduate Center; David Allen, Oregon State Division of Employment; Ted Calouri, Ronald C. Cease, Delna Jones, Oregon State House of Representatives; Jane Cease, Nancy Ryles, Oregon State

Senate; Charles Humble, Steve Jenning, the *Oregonian*; William Glasgow, Perkins, Coie, Stone, Olsen, and Williams; Dolores Winningstad, philanthropist; Brian Dolinar, Jim Hurd, Planar Systems; Elizabeth J. Radigan, Portland Development Commission; Joseph C. Blumel, Lee W. Casperson, Lewis Goslin, Portland State University; Barbara C. Peeples, Public Relations Counsel; George McElmurry, Rogers Oregon Company; Casey Powell, Sequent Computer Systems; Ralph Shaw, Shaw Venture Partners; Wayne Atteberry, Standard Insurance; Gary Conkling, Susan Stone, Tektronix; William B. Webber, Tom Williams, Tektronix Foundation; Louis Pengue, TriQuint; Richard J. Detwiler, Tualatin Valley Economic Development Corporation; Jim Buchanan, The Waggener Group; Wes Myllenbeck, Washington County Board of Commissioners; Joan H. Smith, Washington County Historical Society; and Richard Matthews, Washington County Museum. We are indebted to Marshall M. Lee of Pacific University, the pioneer historian of Washington County's high technology, for allowing us to consult his definitive history of Tektronix before its publication. Above all, we are indebted to Linda S. Dodds, who advised us in the techniques of interviewing and who conducted many of the interviews. Without her aid we could not have completed this book.

The information gained from all of these helpful sources would be as nothing if we had not had indispensable assistance in conveying it to our readers. Philippa Brunsman and Carolyn M. Buan were magnificent editors who performed the duties of their craft with skill and sympathy. Elaine Jolliff transformed authors' and editors' scrawls and scribblings to a finished manuscript. We are most grateful to them.

Finally, a grant to the Oregon Graduate Institute made this study possible. To the donor, Paul McGilvray, we extend our deepest appreciation.

PART 1

THE COMPANIES

1

Beginnings

DURING THE EARLY YEARS of the nineteenth century, and stretching back in time, the lush forests and gentle plains of Oregon's Tualatin Valley were the home of roaming native peoples. By the mid-1800s, they had been joined by white settlers from the eastern United States, who were drawn by the valley's rich soil and mild climate. Throughout the rest of the century, and well into the next, the land that became Washington County was rural, its economy based primarily on agriculture.

With the end of the Second World War all of that changed, as a rapidly growing population spilled over from the city of Portland and a new industry—later known as high technology—looked for a new land where it could build larger plants to replace its cramped urban quarters. In only a few years, high technology's move to Washington County launched the sleepy area into the fast-paced modern world and made it the seat of some of the country's most exciting and far-reaching engineering discoveries. For the most part, change was accomplished in an orderly manner, and the new companies proved to be good neighbors, who respected the pleasant way of life and pastoral beauty that the county offered. In the best sense, it was a marriage of opposites that worked.

The first people in the land that became Washington County—

and later, Oregon's "Silicon Forest"—were the Tualatin, who gave their name (corrupted from "Atfalati") to the 300 square-mile valley that was their home.

The Tualatin Valley consists of a broad plain surrounded by long, sloping terraces. Only two hills of any size (the 750-foot-high Bull and Cooper mountains) rise above the prairie with its scattered groves of trees. It is a landscape kept relatively free from dense stands of forest by the Atfalati's practice of systematically burning the terrain to drive game during their periodic hunts. Draining the valley is the sixty-mile-long Tualatin River, which furnished fish and shellfish and a mode of transportation for the approximately two thousand men and women who made up the Atfalati bands.

Like all Native Americans of the Oregon Country, the Atfalati based their economy on nature's gifts. They gathered camas bulbs, wapato roots, hazelnuts, acorns, and assorted berries; hunted elk, deer, rabbits, and squirrels; and fished for salmon, steelhead, and shellfish. What the Tualatin Valley did not produce, the Atfalati secured by trading with others at the falls of the Willamette and at the great intertribal market of the Columbia River at what is now The Dalles. They also manufactured clothing, necklaces, bows, arrows, and spears for trade and for their own use.

Nature provided the Atfalati with a host of myths and legends to explain the birth of the universe, the origins of men and women, and proper human behavior. Although their culture was complex, it was stable. Material and cultural success for individual and group depended on carefully repeating the tasks, ideas, and institutions sanctioned by the wisdom of the ages.[1] However, the Atfalati's stability was shaken and quickly destroyed by the arrival of white settlers.[2] The Atfalati first encountered white men in the beaver trade, when they sold pelts to the North West Company, the Pacific Fur Company, and the Hudson's Bay Company. In 1840 three former mountain men, Joseph Meek, Caleb Wilkins, and Robert "Doc" Newell, with their Native American wives, retired from the dangerous trapping trade and began farming in the north plains of the valley. Soon an agricultural colony grew up around them and missionaries came to preach the Christian message to pioneer farmers and Native Americans alike.

The first organized party of farmers came to the Pacific Northwest in 1841 and over the next two years hundreds followed. So many came to the Oregon Country that they quickly felt the need for an organized government to protect their lives and property. In 1843, therefore, settlers and missionaries met at Champoeg and created a provisional government. In that same year, four counties were established, among them Tuality, an area extending north of the Yamhill River and west from the Willamette River to the Pacific Ocean. In 1844 Clatsop County was cut off from Tuality, and in 1849 the territorial legislature changed the name of Tuality County to Washington County, in honor of the first president.

By 1851 the number of Atfalati had been reduced by smallpox, venereal disease, malaria, and respiratory complaints to a mere sixty-five. The farmers, on the other hand, survived. In the abundant land, wheat flourished and the only economic problem was to get it to market in an area where roads were poor and floods common. Occasionally the Hudson's Bay Company at Fort Vancouver would take a few bushels, or some could be sold to arriving pioneers; but for the most part the early settlers found that their efforts yielded them only a subsistence living, not a cash crop. However, in 1848 the situation changed dramatically with the news that gold had been discovered in California.

In January 1848 James Marshall found gold on the American Fork of the Sacramento River, and during the year that followed at least eighty thousand men and women from all over the world rushed to the area to make their fortunes mining alongside him. Although many Oregonians were among them, Oregon benefited from California's gold in a more prosaic but ultimately more substantial and enduring manner, by providing food and shelter for the miners. Portland soon became the main shipping port and commercial center of Oregon, and in 1851 its businessmen began making the Canyon Road over the West Hills to the Tualatin Valley. From the road's completion in 1856, the city and the valley reinforced each other's economic growth and, together, brought Oregon out of the despair and backwardness of a subsistence economy.[3]

In 1883, the second great economic turning point in Oregon history occurred when the Northern Pacific Railway, the first transcontinental to reach the Pacific Northwest, arrived in Portland. The railroad's effect on the region was twofold: it made the area an even greater exporter of wheat and other agricultural products than it had been earlier, and it brought many new farmers from the Middle West to the valley.

After the Civil War, and especially after the railroad's arrival, another type of newcomer—the foreign migrant—also began to settle in the county. There, Leahys, Siegenthalers, Hickethiers, and Murrays represented the Irish, Swiss, German, and Scottish migrants, who were joined after 1900 by a sizable contingent of Japanese.[4] By the turn of the century all the good land had been taken, and the only remaining employment was in logging, construction, and city jobs.

Citizens of Washington County remained tied to a rural economy. Its residents raised crops or set themselves up in businesses or professions in the small towns that were growing up to serve the farmers. By 1940, on the eve of America's entry into the Second World War, the county's population numbered only 39,194, with 3,747 souls in the county seat of Hillsboro and 1,052 in Beaverton. Within two decades of the war's end, however, the national economy had burgeoned, and Oregon's population had begun to increase rapidly. Of the newcomers, many found work in Portland but preferred to live in the suburbs of eastern Washington County, which soon came to be known as "bedroom" communities. To accommodate the growing population, the old Canyon Road was converted into a freeway and housing developments began to replace farms and orchards. School districts consolidated and new school buildings were built. Inevitably, problems associated with government and social services became more intricate. By 1960 the population of Washington County had risen to 92,237 and that of Beaverton to 5,937.

In part, this growth was the by-product of the new type of industry known as high technology, which was represented by companies such as Tektronix, a firm that had moved to Washington County in 1951.

HIGH TECH: ORIGINS OF AN INDUSTRY

The term "high technology" was not coined until the 1970s and came to mean different things to different people. For our purposes, however, a high-technology industry can be characterized as one with a work force of skilled employees, many of whom are scientists and engineers; a high ratio of research-and-development expenditures to sales; a rapid rate of growth; and a worldwide market for its products. As we shall see, these industries produce diverse and extremely technical products, which cannot be neatly or briefly defined.

An early advance along the United States' road to high technology occurred in 1906, when Lee De Forest of California invented the vacuum tube, a device to detect and amplify electrical signals.[5] In 1912 he discovered the vacuum tube's amplifying qualities, an insight that made possible the development of radio (and later television) and led to the formation of great corporations such as the Radio Corporation of America and General Electric. Hewlett-Packard, the first major electronics firm on the West Coast, was created in 1938 by David Packard and William Hewlett to build variable-frequency oscillators, devices originally used to diagnose problems in electrical equipment, which were refined after the Second World War for use in high technology.

In 1947 William Shockley, John Bardeen, and Walter Brattain, of the Bell Telephone Laboratories, invented the transistor, a solid piece of material that "allowed the magnification of electronic messages, as did vacuum tubes, but required much less current, did not generate as much heat, and was much smaller."[6] They found that they were able to amplify an electronic signal by passing it through a semiconductor (a "solid-state" device). According to one definition:

> A transistor is a *semiconductor*, a class of materials whose ability to conduct electricity lies between that of insulators, like rubber, and conductors, like copper. Silicon, the chief component of sand, is the most plentiful semiconductor, followed by such exotic materials as germanium, selenium, gallium arsenide, and lead sulfide.[7]

In 1952 Bell Telephone first placed transistors in commercial production—in hearing-aid devices.

In 1955 Shockley founded the first semiconductor firm, Shockley Semiconductor Laboratory, at Palo Alto, California. By 1957 some of his employees had set out on their own to launch Fairchild Semiconductor, the first company to use "silicon transistors whose circuitry could be placed on a transistor by diffusion through openings in silicon oxide."[8] In 1959 Jack Kilby of Texas Instruments and Robert Noyce of Fairchild Semiconductor, working independently of one another, invented the integrated circuit, "a type of semiconductor that incorporated several transistors—ten or twelve then, up to half a million now—on a single piece of silicon."[9] Soon the transistor and the integrated circuit were coupled with the greatest symbol of the age of high technology, the computer.

If one adopts a broad enough definition, the term "computer" may be traced to antiquity, but the first digital computer was invented by Howard Aiken and constructed by IBM, beginning in 1944. The following year the first electronic computer, ENIAC, went into use. Built at the University of Pennsylvania, the machine contained about 18,000 vacuum tubes. Like other early computers, ENIAC had limited storage, a problem not solved until J. Presper Eckert and John W. Mauchly (its designers) and John von Neumann invented the stored-program computer. UNIVAC, the first successful machine of the type, was put into use by the U.S. Bureau of the Census in 1951. Like all the early computers, the UNIVAC was a large "mainframe" machine operating on vacuum tubes.[10]

The computer provided a large potential market for the transistor and an even larger one for the integrated circuit that followed. Writing in *Scientific American*, Noyce explained later:

> The reason is that digital systems require very large numbers of active circuits compared with systems having analogue amplification, such as radios. . . . An analogue system. . . . cannot handle large numbers of microcircuits, whereas a digital system requires them; a pocket calculator contains 100 times as many transistors as a radio or television receiver.[11]

In 1968 Noyce and another Fairchild employee founded Intel Corporation, where three years later an engineer named Ted Hoff invented the microprocessor, "a semiconductor chip that serves as the central processing unit controlling a computer."[12] This invention, in turn, led to the microcomputer industry—the manufacturing of personal computers for a mass market, as well as for telecommunications, computer-aided design, engineering, and manufacturing.

The story of high technology in Washington County is a story of the continuation of trends begun elsewhere. Equally important, it is an account of new developments that have had their origins here. In both cases, the industry has had an enormous influence on the county. However, Washington County's high-technology industry began not in the Tualatin Valley in recent years, but in Southeast Portland in the thirties and forties.

2

Electro Scientific Industries

OREGON'S ORIGINAL "SILICON FOREST" was on Portland's Southeast Hawthorne Boulevard, in Douglas Strain's company that eventually became Electro Scientific Industries and in Jack Murdock and Howard Vollum's appliance store that later became Tektronix. Both of these high-technology firms were rooted in the U.S. Forest Service Radio Development Laboratory. It is this institution that is the earliest source for Washington County's oldest high-tech company, Electro Scientific Industries.

Region Six of the U.S. Forest Service includes some of the most rugged terrain in the continental United States, making communication by road or trail difficult and sometimes impossible. In the early days of the service, emergencies required the use of a new technology—radio. Moreover, since the technology was new, the Service was interested in improvements and assigned some of its mechanically adept employees to its Radio Development Laboratory, where they not only repaired existing radios but produced new equipment. In time, they became skilled enough to make rather elaborate portable battery-operated units. By 1931 this manufacturing enterprise (situated first in Tacoma and later in Vancouver, Washington) was now encroaching on space required for other shop work. The regional forester therefore acquired a building on 122nd Avenue and Glisan Street in Northeast Portland—fittingly enough, from radio station KEX, whose original transmitting towers dominated the grounds.[1]

The most notable employee of the laboratory was Harold Lawson, the real pioneer of Oregon's electronics business. Lawson was born in 1904 and became interested in the new device of radio as a boy of twelve living in Vancouver, Washington. There, he learned to build his first receiving set from the local library's two books on the subject. After the First World War he acquired his first vacuum tube from an army radio-telephone operator who was stationed at Fort Vancouver.

As a young man Lawson studied engineering at Oregon Agricultural College (now Oregon State University) but left without taking a degree. After stints as a straw boss in a logging camp and a partner in an electrical-contracting business, he joined the Forest Service.[2] At its laboratory, he designed a new radio transmitter, the "M" set, which proved the value of radio communications in fire control. His knowledge made him the senior technician in this area and later, the respected mentor of such important figures in high technology as Howard Vollum and Douglas Strain.

Strain's connection with Lawson and his first contact with Oregon came during the Second World War.[3] At this time, following the attack on Pearl Harbor, there was widespread fear that the Japanese would use submarines to bomb and shell the Pacific Coast. Particularly dreaded were balloon-directed fire bombs that could ignite the Northwest's forests, which were of critical importance to the war effort. A natural place to develop a defense against such threats was the Forest Service Radio Laboratory, to which Douglas Strain was assigned. The choice of Strain was particularly apt.[4]

DOUGLAS STRAIN, FRANK BROWN, AND THE BROWN ENGINEERING COMPANY

Strain was born in Spokane but spent most of his boyhood in Idaho, where he built his first receiving set around a Quaker Oats box so that he could listen to station KSL in Salt Lake City. Later, during his high school years in Pasadena, California, he deepened his interest in physics by attending the free Friday-night lectures at the California Institute of Technology given by such luminaries as Robert A. Millikan, winner of the 1923 Nobel Prize in physics.[5]

While he was in high school, Strain worked part time installing radios in automobiles, and then on the assembly line and in the engineering laboratory of a radio-manufacturing firm. After two years at Pasadena City College he transferred to Caltech but was caught up in the war before finishing his degree. He spent his wartime service first at the University of Illinois, where he did neurological research and built neurological simulators, and later in Portland, at the Forest Service Laboratory. After the war, he completed his degree at Caltech and worked for a time on spectrophotometers and pH meters at Beckman Instruments in Pasadena, returning to Portland in 1949. Part of his reason for moving was that he and his wife, Leila, did not relish raising their children in the "rat race" of Southern California.[6]

The person who brought Strain north was Frank Brown, the founder of Brown Engineering Company of Portland. The story of Brown's career is both interesting and significant to the development of high technology.

One of the major early radio-manufacturing companies in the United States was General Radio (now GenRad), a Massachusetts firm whose first manufacturing plant was at Walden Pond. The founder of General Radio, Melville Eastham, had been raised in Oregon at Klamath Falls, where he had known Brown. During the Second World War, General Radio secured a contract to build an impedance bridge, an instrument for measuring alternating-current resistance, which it had invented in 1933. (The impedance bridge is similar to the Wheatstone bridge, which at one time was familiar to all high school physics students.) When the company needed additional help to fulfill its contract, Eastham remembered Brown and subcontracted part of the work to his Brown Engineering Company. In turn, Brown sometimes asked for help from the technicians and engineers at the Forest Service Laboratory, and it was there that he first met Strain. In 1949 Brown and another engineer invited Strain to become a partner in Brown Engineering, which even after the war continued to manufacture impedance bridges. Strain served as president for research and development in the firm—later known as Brown Electro-measurements Company, or BECO.

ESI'S EARLY YEARS

In 1953 Douglas Strain and his father, Clayton, bought out his other partners' interests in the company for less than twenty thousand dollars—a move significant to the history of high technology in Oregon. They had some difficulty in naming the company. Their first choice was Electro Measurements, Inc., abbreviated to EMI, but an English firm, Electrical Musical Industries, had registered a trademark on the initials. The Strains then decided on ESI (which didn't stand for anything) simply because they liked the ring of the three letters. The name lasted a year, until Clayton Strain complained. As Douglas remembers, "Dad didn't like 'ESI' because nobody had paid their bills. They kept writing saying, 'We have your invoice, but what's your name?'" The name was changed to Electro Scientific Industries in December 1959. It was typical of the founders' philosophy that the company was not named for individuals.[7]

Another difficulty was that the father and son had to raise the entire purchase price from their own assets, even though they owned a company producing a high-quality product that was in demand. Local banks were unwilling to lend money in the new field of electronics, a reflection of the Portland financial community's general conservatism. Oriented to the old economic style of lumber products, Douglas Strain recalls, "they would loan money on 2 x 4s, but they wouldn't loan money on capacitors or vacuum tubes." It was impossible to borrow even on electronics inventories or accounts receivable, and that was years before venture capital became available. Besides being unwilling to participate in endeavors that lay outside their own experience, Portland bankers had no knowledge of successful electronics loans in other high-technology centers such as Boston or San Francisco. Not until

1958 did a young banker from U.S. Bank—John Elorriaga, later its retired chairman—persuade a Portland bank to make its first loan to ESI.[8]

Undeterred by this conservatism, Strain pushed ahead with his new company. His three top managers were all from California. Clayton Strain, the secretary-treasurer, had graduated from Oregon Agricultural College in dairy manufacturing and had pursued graduate studies in economics at Cornell University. He had held offices in local government and had worked in the creamery business for twenty-eight years, the last twelve as credit manager of the Challenge Cream and Butter Association of Los Angeles. Both the other managers had known Douglas Strain at Caltech; one of them, Lawrence B. Rockwood, had worked with Strain at Beckman Instruments, where Rockwood was chief engineer and director of research. He later resigned to become vice-president for manufacturing in the new firm, while Merle Morgan accepted Strain's offer to become director of research and engineering. Morgan had received his doctorate from Caltech in 1954 for his dissertation on the design of a computer that could perform complex algebraic functions. Like Strain, Rockwood and Morgan were attracted to ESI not only because they found it challenging but because Portland and Oregon appealed to them and their families.[9]

The original number of employees at ESI was twenty-four, a work force largely comprising women, who did the manual wiring and soldering handwork required in those pre-automation days. Some of them had been trained by a local firm, Teeples Electronics, which they had left when it had been sold to an out-of-town company. Others were trained at ESI. The plant, a forty-by-sixty-foot building, still stands at Forty-seventh Avenue and Hawthorne Street in Southeast Portland.[10]

For about seven years, labor and management worked to lay the foundation of a successful company, continuing to manufacture an improved universal impedance bridge (Model 250). The company's main market was still, and would continue for many years to be, the Department of Defense. However, as other precision instruments went into production, it became necessary to expand the marketing network. The company was too small to have its own sales force in the United States, let alone abroad, but, as in the past, California came to the rescue—this time in the form of Hewlett-Packard. That company already had an international reputation, with sales representatives around the United States and half a dozen organizations in Europe that sold on commission. These facilities were offered to electronics firms that were not competitive. In the Pacific Northwest, the first company to accept was Tektronix, followed by ESI. Close cooperation among the three firms later led Hewlett-Packard to offer to buy out Tektronix and ESI, but both Oregon firms preferred to maintain their independence. Yet the Hewlett-Packard sales connection had been indispensable to ESI. When the company had begun operations, there were only three other manufacturers of impedance bridges in the world: one on the East Coast, one in England, and one in Germany. Hewlett-Packard provided the international market access and sales force that ESI needed from the beginning.[11]

The products underlying the company's success from 1953 to 1959 ranged widely. In addition to the universal impedance bridge, there were the precision voltage dividers, known as "Dekopots," which Frank Brown had originally put on the market. In 1954 ESI began selling a new type of analog computer, the ESIAC, which Morgan had designed to obtain rapid solutions to complex algebraic functions. The National Aeronautics and Space Administration bought it to redesign the stability of its booster rockets, which had been blowing up on their launching pads. The Boeing Company also used it to design its 707 jets. Before the invention of digital computers destroyed the market for analog computers, ESI sold about sixty ESIACs. As Douglas Strain remembered: "We had visions of putting one of those on every design engineer's desk at $10,000 a throw, but it didn't happen."[12]

In spite of skilled management, innovative products, a sales force, and international markets, ESI's first years were lean and managers were often paid in shares of stock. (Douglas Strain recalls that at times soup stock would have been preferable to company stock.) However, by early 1957 the future looked brighter and the company moved to a new factory at 7514 Southwest Macadam Avenue. On 18 July that year the entire engineering and production facilities burned to the ground, providing a spectacle for the Pacific and Intermountain Fire Chiefs, who were gathered in Portland for their annual meeting. By the end of the day only the offices remained intact, yet within another day the company was back in business, scattered in four locations around the city. Aid was extended by various firms, including Tektronix, but the financial blow was grievous, and it took several years for the company to recover fully from the disaster. In 1958 a new air-conditioned plant was built at Macadam Avenue and Spring Garden Street, near the Sellwood Bridge.[13]

With the formative years came a corporate style, consisting, in Strain's words, of "strong support for education, liberal treatment of the employees, enlightened personnel policies, lack of ostentation in management, lack of hierarchical structures. . . . "[14] This style came from his own character and outlook and from other high-technology firms during the industry's early years in both California and Oregon. Their approach represented an updating and improvement of the "welfare capitalism" of the twenties, which promised employees a job that was interesting, well paid, and so satisfying in all respects that they would not want assistance from unions or government. Strain credits the influence of Hewlett-Packard and Tektronix in developing the corporate philosophy of ESI and regards it as fortunate that the firms were so influential in the West Coast high-technology industry.

From the outset, Strain envisioned an ideal for his firm—as did Jack Murdock and Howard Vollum for Tektronix—that was more than material profit. He talked of service as the goal of his firm. He knew that nonprofit organizations—schools, churches, symphonies, historical societies—were widely regarded as serving their communities and their country, and he hoped his employees would also see their work as having wider goals than service to customers and stockholders. For his employees, he "tried to put the concept of profit in a little different focus. It's really not the ends of business, as we stated in our principles—it's the means that make the proper ends possible. . . . People didn't feel they were just, shall we say, working for the almighty dollar. There were a lot of things we were able to do as a company that really did help advance science and helped advance our community." Strain took many of his statements of purpose from Hewlett-Packard and Tektronix manuals, but was also influenced in his open-management style by Douglas McGregor, a teacher at Antioch College (and later at MIT) and in 1960 the author of a book called *The Human Side of Enterprise*.[15]

THE YEARS OF EXPANSION

The company opened the sixties with a new name. "Electro Scientific Industries" was chosen as being an accurate description of the company's interests in a variety of electronics products. It was an apt change and a timely one, for the company was about to attain greater diversity, profits, and recognition. Behind the new emphasis on electronics was the sputnik.

The first artificial satellite, lofted by the Russians in the fall of 1957, set off a tremendous boom in technology and science in the United States, as the nation tried to catch up with and surpass the Russians. Colleges, universities, high schools, and private business were all brought into the effort to boost American science and technology. For ESI this alliance took the form of many new contracts with the Department of Defense. In 1958 the company supplied instruments to General Dynamics (Convair) for the Atlas Intercontinental Ballistic Missile Program. Unmanned satellites and the manned space program also required precise measuring instruments in military, university, and corporate laboratories—instruments that ESI could provide.

In 1960 the company signed a contract with the U.S. Air Force to develop more precise electrical measurement instruments. Although other major electronics firms had had preliminary contracts for the purpose, the small Portland firm was the winner in the final competition, with its Model 242 Precision Resistance Measuring System. Later it bid on other components of the program and won eighteen of the twenty contracts offered. Deliveries of the equipment began in 1963 and were completed the next year, without a single piece of equipment rejected in the Air Force acceptance tests. The $1 million project contributed to the equipping of 160 Air Force precision measurement laboratories. Even more significant was the sale of ESI instruments to civilian laboratories all over the world.[16]

Success with the Model 242 also led to a move to new manufacturing facilities. Originally the company had intended to acquire new land next to its Macadam Avenue plant, but the City of Portland had placed options on all the adjacent land for the development of water facilities. Other sites in the city were investigated without success. Portland gave ESI the impression of not needing or wanting new light industry within the city limits.

For ESI a way out of this stifling situation had been shown by Tektronix, which in 1951 had outgrown its Portland premises and had moved to Washington County. In 1962 ESI decided to go a step further and bought one hundred acres near the junction of Murray Road and the Sunset Highway (Highway 26). (It was not the company's first interest in Washington County, however, for in 1959 Clayton Strain had taken an option on forty acres near Cornelius as a possible plant site.)

The Sunset project was based on the success of the Stanford University Science Park, created in 1951 by the university's vice-president, Frederick Terman. ESI's Sunset Science Park was the first in the Pacific Northwest designed to attract light manufacturing industries interested in research and development to an environment modeled on a university campus.[17] Although the "park" was a regional innovation, it was not originally welcomed with open arms by local citizens and politicians. The county had little industry, and residents were dubious about putting an industrial plant close to housing developments. Parents of children at the new Sunset High School and residents of the new Oak Hills housing development were especially unhappy, making no distinction between light industry and lumber mills, between clean plants and spewing smokestacks. It took much persuasion by company officials, the intervention of Gov. Mark O. Hatfield (who was also supporting the infant Oregon Graduate Center), and the urging of the architectural firm Broome, Oringdulph (who had designed Oak Hills) to get the area zoned for a high-technology park.

In 1963 ESI established offices in the new area, with Governor Hatfield presiding at the dedication ceremonies.[18] Three years later the company moved all its operations to Sunset Park. Its new manufacturing plant, of prefabricated concrete-well construction, covered thirty-six thousand square feet with another six thousand square feet located on the mezzanine floor for cafeteria and other facilities. In 1980 the company bought the adjacent Sunset Valley Grade School and moved its administrative offices there, in the process acquiring the school gymnasium, which became popular among the employees.[19]

Although the move to Washington County was an almost unqualified success, Strain's ideal failed in one respect. He had anticipated that ESI employees would live near the company—in Oak Hills or other developments where they could walk to their work, which in turn would be near schools, banks, and service businesses. Later, he concluded ruefully that the plan never caught on: "We all tend to live in different places and drive to work, regardless of the opportunity of living close by our work, which was the general idea."[20]

During the sixties ESI pursued other new directions. The company had gained an excellent reputation with its impedance bridges and Air Force contracts, making it easier to find outside financing for its products. All of its new operations, only some of which were successful, were built on its impedance-bridge base. In 1964, when the original Air Force project was completed, the company hoped to sell space equipment to the Air Force and expand into naval vessel and army mobile applications. But these hopes were dashed when the federal government diverted military funds to other uses during the Vietnam War, a diversion that cost the company 40 percent of its shipments and caused a drop in 1964 and 1965 profits.[21]

The new ventures of the sixties included the creation of ESI's first subsidiary, Mikros, which was to manufacture a new type of electron microscope based on electrostatic, rather than the more widely accepted electromagnetic, principles. Mikros also made power supplies, an automatically programmed vacuum evaporator for the optical industry, and a cardiac defibrillator. In 1964 the subsidiary was sold to Varian Associates of Palo Alto; proceeds from the sale were invested in the development of a new line of voltage- and current-measurement equipment. Chief among these was the Model SR104 Transportable Resistance Standard introduced in 1966. The Model SR104 became the most popular standard of its type throughout the world, gaining acceptance with national bureaus of standards in several nations.

By the end of the sixties ESI was an internationally recognized firm, whose major customers included the U.S. government, universities, Boeing, Westinghouse, and General Electric. In the middle of the decade, however, the company had to change direction rapidly. The early age of electronics, which had emphasized research, was ending; the new era of production and big business in electronics was beginning. ESI made the transition magnificently—through the laser.[22]

ESI ENTERS A NEW ERA

The seventies saw ESI's first interest in the laser, a narrow, intense beam of light with a myriad of uses. The company became a pioneer in the use of lasers by having a computer direct its instruments and using the laser beam to burn away, or trim the impurities in its circuits. The laser soon became the heart of the firm's program and its ticket to the future. It made ESI much larger and more diversified and necessitated its becoming a public company in the eighties. The answer to why ESI began the development of systems based on the laser lay with the ceramic or silicon chips that are the basis of modern electronics.

In the manufacture of integrated electronic microcircuits (the "computer on a chip"), a thin film is deposited on them in a pattern that both connects the circuit elements and forms the number of resistors. The value of the resistors determines the performance of the circuits. Ordinary techniques of depositing film do not allow the precision necessary to ensure the best performance. Rather, the film must be "trimmed" until the desired resistance is reached. Originally, air and sand abrasion were used to trim circuits, but the laser proved to be faster, cleaner, and more precise than these techniques.

ESI had become involved in resistor trimming in the late sixties. Working in cooperation with General Electric in the early stages of research and development, ESI produced its first laser-trimming system, Model 44, in 1976. Model 44 subsequently became the most popular laser trimmer ever developed and was supplied to virtually every manufacturer of microcircuits in the world. Television sets, automobile ignitions and fuel injection systems, telecommunications—almost all electronics projects use microcircuits. ESI's success would be assured if it could keep pace with advances in microcircuitry, especially the shrinking of semiconductor devices to minute dimensions.[23]

The company's pace was rapid. In 1978 the first successful redundancy-memory "link-cutting" experiments were completed with a modified laser-trimming system and an experimental short-pulse laser. In 1979 came the introduction of a four-position parts handler for laser systems, which greatly improved speed. The next year saw improved optics and lasers for the semiconductor market. This product opened the massive, worldwide, and potentially unlimited semiconductor market to ESI.[24]

In 1982 the Model 80 was improved with an alignment subsystem that permitted the laser beam to be moved more accurately to other positions on the same die. The next year it was further enhanced with the addition of an automatic wafer handler that permitted cassettes of up to fifty wafers to be processed without an operator's attention. In 1984 came ESI's Model 8000 Laser Processing System, a third-generation system. With these advances in hardware, which had been occurring since 1970, came greater computer control with the introduction of Pascal software at ESI in 1976, when the Model 44 was brought in. All these developments made the newer, more complex circuits packed on each wafer more reliable and hence more valuable in volume production.

The financial success of the new laser products was impressive. From 1976 to 1979 sales doubled, and by 1980 had reached $33 million compared with $21.9 million in 1979.[25]

FOREIGN SUBSIDIARIES AND PUBLIC FINANCING

To finance these advances in laser-use technology, ESI moved into new areas, among which was the development of wholly owned foreign subsidiaries. The principal advocate of this venture was Paul D. Lintner, vice president of marketing, who became president of the company in 1979.

ESI had been involved with many other countries since it had begun developing resistance standards. With laser systems, a world presence was imperative for the company whose customers and potential customers—for example, Siemens, AT&T, and IBM—had facilities all over the world. ESI had to deal with them on the spot rather than by communicating back to Oregon.

The first subsidiary, which was the foundation for successful sales and service operations all over Europe, was ESI, Gmbtt. It opened in Munich in 1977. The second, ESI, K.K., established in 1981 in Tokyo, reached the Asian markets. Successive foreign subsidiaries—ESI, S.A.R.L. in Paris, and ESI, Ltd. in London—opened in 1982.[26]

Also in 1982, ESI acquired its first corporate U.S. subsidiary after Mikros, Palomar Systems and Machines, Inc. of Escondido, California. It was an arrangement suggested by customers of the two companies who were chip-component manufacturers and who wanted a single system to bring together Palomar's handling equipment and ESI's microprocessor-based instrumentation in order to increase productivity. The acquisition involved the exchange of 1.36 shares of Palomar for each share of ESI.

At the time of the acquisition, Palomar employed forty-five persons and had annual sales of about $4 million, while ESI had six hundred employees who achieved an annual sales of $35 million. The acquisition of Palomar provided ESI with two advantages: it helped the company enter new product areas two to four years earlier than would have been possible on its own, and increased sales, with Palomar contributing 60 percent of ESI's sales in 1982. A second acquisition followed in 1986, when ESI acquired Metoramic Sciences of

Carlsbad, California. The three combined firms offered complete production capability to passive-component manufacturers. ESI supplied the instrumentation, Palomar the handling equipment, and Metoramic Sciences the raw materials of inks and binders.[27]

By that time, the volume and complexity of ESI's operations demanded changes in its financing. For the impedance bridge products, financing had been secured from New York, from San Francisco, and from the State Farm Insurance Company of Bloomington, Illinois. In 1982 two private placements of stock were completed through F. Eberstadt and Company of New York. Shares were sold both by the company and by individual stockholders. After the two transactions, approximately 12 percent of ESI common stock was held by the Eberstadt group. It was hoped that this acquisition of capital would permit the company to grow substantially before additional sources of funds would be needed. However, the hope soon proved to be false, and in 1983 ESI joined Tektronix and Floating Point as Oregon's principal public high-technology corporations. The step was taken with some reluctance by a company that had always relished its freedom as a private entity, but the company had become so successful that it was no longer possible to finance expansion with earnings or with the type of borrowing it had formerly employed. In 1983 the stock market was stable and provided an opportunity to get outside capital at a reasonable price. In August, therefore, ESI went public, filing with the Securities and Exchange Commission an initial combination public offering of 1.2 million common shares. Of that total, 840,000 shares were to be issued by the company and 360,000 were to be sold by certain holders. The lead underwriters for the offering were F. Eberstadt and Hambrecht and Quist.[28]

ESI'S EMPLOYEE RELATIONS AND REPUTATION

In 1981 ESI institutionalized the approach that perhaps had most contributed to its success: its almost unlimited willingness to draw on its employees' intellects and creativity. In that year it created its PAL (People Action Links) program. Since the company's inception, Douglas Strain had insisted that employees be allowed to participate in its running. Stock-option plans and profit sharing came early in its history as obvious manifestations of the founder's concern to motivate his employees. Strain had insisted that the managers listen to the employees, not only formally but sincerely. "Every new idea," he said, "was once a minority idea by definition." Veteran employees recall the human, familial spirit of the company from its early days. As the company grew larger it was not so easy to give everyone a hearing, but PAL provided a way to continue involving employees in the management of a large company. Through PAL, groups of employees decided what project to work on or what problem to solve, calling on the various company resources for help in planning their proposals, which were then presented to management for a final decision.[29]

ESI's employee-relations policies have had other unusual approaches. In the early sixties a concerted effort was made to bring black workers into the electronics industry. Several electronics firms, including ESI and Tektronix, cooperated to establish the Northwest Electronics Workshop. A training station was set up on Union Avenue in Northeast Portland, where prospective employees were given real jobs assembling components. The plan was an idealistic experiment that didn't fully succeed, mainly because the newly trained employees lived too far from the manufacturing plant, on the other side of Portland. More successful was the "sheltered workshop" program, which gave work to handicapped persons through contracts to the firms, or workshops, that employed them. Some of the program's employees later became permanent workers for ESI. The company was also a leader in hiring handicapped workers directly.[30]

National recognition for ESI's labor policies and programs came in 1983, when the company was listed in the book *The 100 Best Companies to Work For in America*. The authors chose their "best companies" on the basis of five criteria: pay, benefits, job security, chance for advancement, and ambience. ESI was rated "superior" in job security and advancement, and its belief in employee self-management was singled out for praise. The outside evaluation was spectacular confirmation that ESI was living up to Douglas Strain's principles of management: to manage the business with the primary objectives of making a contribution to society; to recognize the dignity and personal worth of every individual; to recognize a responsibility to the community; and to develop and encourage a better understanding of the nature of profit. It was a fitting tribute to a man who would soon retire (in October 1985) from the company that—with Tektronix and the Radio Development Laboratory—had laid the foundation for Oregon's high-technology business.[31]

3

Tektronix

MURDOCK AND VOLLUM; VOLLUM AND MURDOCK: the two surnames conjure up not only a company and a product, but an environment, a way of being, and a sense of optimism.

In the sixties and early seventies most Oregonians were vaguely aware of Tektronix as a large company, a looming presence that employed an army of workers, boosted the local economy, and brought attention to Oregon. By the eighties, moreover, the firm had become a bellwether of business in Oregon and the trumpeted savior of the state's economy following the collapse of the timber business in 1979. Tek was Oregon's largest private employer and its industrial leader. If Tek was not making money, if the headlines warned of more layoffs there, things must be tough all over.

But prepotency had been hard won.

In 1945, after the end of the Second World War, Jack Murdock and Howard Vollum, with Miles Tippery and Glenn McDowell, formed a company known originally as Tekrad, and later as Tektronix.

In a real sense, however, Tektronix began before the war, when Vollum, a 1936 physics graduate of Portland's Reed College, met Murdock, who had been operating Murdock Radio and Appliance at

Southeast Sixty-seventh Avenue and Foster Road since 1934. The two young men became friends, and eventually Vollum rented space in Murdock's store for a radio-repair business.

Vollum had long been interested in electronics and had built his own oscilloscope by the time he entered college. The oscilloscope—or oscillograph, as it was originally known—was a device for measuring and displaying electrical impulses on a Cathode ray tube, which had been invented shortly after the First World War. In the thirties, however, it was still a rather primitive instrument, used mostly by engineers to diagnose problems in electrical equipment stemming from faulty wiring, resistance, and circuitry.

While Howard Vollum's interest tended toward electronics, Jack Murdock's interests lay in business, specifically in sales. After graduating from high school, he had opened his store with financial backing from his parents. He had a winning personality and a knack for salesmanship that had made his business prosper even during the precarious years of the Depression.

Vollum and Murdock stayed together at Sixty-seventh and Foster until 1940, when Vollum closed his portion of the business and went to work supervising the Radio Project at the National Youth Administration, a New Deal agency. Not long afterward he was drafted into the army and was assigned to the Signal Corps Electronic Training Group, which was then creating a radar capability for the military. In the Signal Corps, he worked throughout the war in what was probably the most sophisticated electronics research of the day.[1] Murdock, meanwhile, continued to operate his store until he joined the Coast Guard, in 1942.

Although the war separated the two friends, Vollum's name came up when Murdock talked to his Coast Guard comrades about starting a postwar business. Among others, Murdock discussed the project with Tippery, a radio repairman.

Murdock and Tippery emerged from their military service intent on starting a new business immediately, but with little idea of what it would be. To some extent, they were limited by available capital. Both men had five thousand dollars in war-bond savings and Murdock could contribute the Sixty-seventh and Foster building, but apart from that, they had little cash.

As it turned out, however, cash was not important, since neither man had great ambitions. Rather, they wanted a small, friendly company of some ten to twelve people, where an informal "family" atmosphere would prevail.[2] In this they were largely guided by Murdock's instincts and long interest in personnel management.[3]

When they finally met with Vollum in late November 1945, he readily agreed to join them and convinced them that they should be in the business of manufacturing oscilloscopes. After adding McDowell, who was skilled in accounting, they were ready to proceed. In early 1946, capitalized by the four partners at $10,400, a company was formed, with Vollum as chairman of the board, president, and chief engineer; Murdock as secretary, vice president, and general manager; McDowell as treasurer; and Tippery as service engineer.[4]

Vollum's interest in the oscilloscope lay in the bright future he saw for it. Despite the ingenuity concentrated on the instrument during the war, the oscilloscope remained a rudimentary device for measuring continuous or repetitive events.[5] However, Vollum believed that after the war, engineers would require a much more reliable and advanced oscilloscope which could measure random as well as continuous events. He was determined to build such an instrument.

In the basement of his parents' home, therefore, he began his research and development, while Murdock and Tippery rounded up, sorted, and stored bargain-rate, government-surplus electronics parts. The initial Tektronix oscilloscope, the 501, took shape on Vollum's workbench in 1946. The "5" stood for the five-inch cathode-ray tube it utilized and the "01" for the first of the line.[6] At eighteen inches wide, about three feet high and three feet long, it was innovative, but too big.[7] With the help of Milt Bave, a gifted machinist and designer whom Murdock had met in the Coast Guard, Vollum began shrinking the instrument to reasonable proportions.

In May 1947 the Series 511 oscilloscope was completed. Significantly smaller than the 501, it represented a revolution in commercial oscilloscopes.[8] At the time, it was the only instrument to feature a calibrated amplifier and calibrated time base.[9] It also had more advanced circuitry than any other oscilloscope, which made it more reliable.[10] In addition, it could measure highly elusive random events.

Marshall Lee, the historian of Tektronix, describes the oscilloscope as follows:

> [It] "writes" a picture of an electrical signal on the screen of a cathode-ray tube. This signal can be a very-low-frequency signal, such as emitted by a microphone as it picks up the sound of a bass violin, or it can be a very-high-frequency signal, such as the electrical output of a short-wave transmitter used for international broadcasting. Engineers and technicians need to know how big signals are, expressed as amplitude and measured in volts. Normally, these signals are repetitive in waveform, and it is important to be able to measure accurately the time interval between equal points on repetitive waveforms.[11]

The ability of Vollum's machine to address these needs with precision and reliability not only "transformed the oscilloscope from a qualitative into a quantitative measuring device" and set it apart from those manufactured by Dumont, RCA, and Varian, the previously dominant U.S. brands, but earned for the 511 the sobriquet the "Vollumscope."[12]

In the first year of their new firm's operation, Murdock and Vollum drew no salary, while together McDowell and the married partners, Tippery and Bave, took out only $1,492.[13] Nevertheless, the company enjoyed a growing sense of solidity. Thanks to the surplus parts that had been gathered, the 511 could be sold at a cheaper price than other oscilloscopes. There was a need for larger quarters. A new building was constructed on the corner of Southeast Seventh and Hawthorne. The young company moved there in January 1947.

THE FOUNDATIONS OF TEK CULTURE

It was during the years at the Hawthorne plant that the seeds of Tek culture were sown. As orders for the 511 started to roll in, it dawned on the partners that their original estimates of company size could not work. By late 1948 they realized that the ideal number of employees might come to as many as eighty.[14] At that number, they speculated, everyone would still know one another on a first-name basis, all could still directly involve themselves in major decisions, and—perhaps most important—the air of equality and informality that was nurtured at the company from the beginning would survive.

At the time, the company's values were inculcated and manifested in a number of ways. The partners based ratios for the directors' salaries on profits rather than allocating them in flat amounts.[15] They eschewed special privileges, such as company cars and first-class travel.[16] While both Vollum and Murdock had strong egos, none of the owners was particularly anxious to be wealthy or in command. This fostered an egalitarian spirit, which in turn reinforced a sense of shared enterprise that was passed down from generation to generation of Tektronix employees.

This outlook was encapsulated in long, often chaotic shareholders meetings, where the partners' personalities were given a forum and topics of discussion ranged far more widely than those that more businesslike companies would have pursued. These meetings became the model for an atmosphere that was contentious yet cooperative.[17]

The benefits of that tone were most certainly felt by the engineers working in the open second floor of the Hawthorne building with its low partitions separating individual workspaces. Like backyard neighbors, the engineers were able to converse, exchange information, and help each other. The enterprise became almost a collective act, and Tektronix engineers who held patents seemed embarrassed at being singled out in such a way.[18]

The company's approach gave free rein to the engineers' confidence and consequently to their ingenuity. For example, an engineer named John Kobbe came up with an innovation called "DC-coupled unblanking," a method of unblanking the cathode-ray tube, the heart of the oscilloscope, into a display mode by sampling the input signal. Before sampling, the screen could remain blank—an improvement over other oscilloscopes, which always had a tracing on the screen, even if no current was being tested. Noting the improvement, one of the company's major competitors observed that he had thought of the concept years before but had never tried it because he knew it wouldn't work.[19] At Tektronix the free exchange of ideas and their critical examination engendered a creativity in the engineers that made what seemed impossible to others appear merely difficult at Tektronix. To gain acceptance of his concept, Kobbe had been forced to sharpen and defend the idea to his superior and to enlist the aid of his fellow engineers. The result was eventually a team effort and a significant technological advancement.

If the free rein given to ideas at Tektronix pushed innovation along, quality remained a primary concern. From the first, there was an obsession with perfection that was itself driven by the pace of invention. As a result, the impulse to build only the finest products was prominent in Tektronix culture. An example of this in the early years centered on the cathode-ray tube.

At the time, Tektronix used cathode-ray tubes manufactured by both RCA and Dumont, but for various reasons the tubes were chronically unreliable.[20] When Tektronix engineers needed tubes built to new specifications, they had to make their needs known to the two suppliers. This presented three problems, one of which was that the suppliers, unhappy about manufacturing new tubes with a limited market, dragged their feet. At the same time, in ordering new tubes, the Tektronix engineers inevitably revealed important information to two companies that were competitors. RCA and Dumont were the world's leading manufacturers of oscilloscopes until Tek captured their markets in the mid-1950s. The orders were evidence of new initiatives and of advanced thinking. Moreover, as innovation at Tektronix accelerated, the standard RCA and Dumont cathode-ray tubes were too primitive for the company's new oscilloscopes. They became not so much components of the instruments as obstacles on the path Vollum and his engineers had chosen for their company.[21]

In 1949, in the face of poor parts, Tektronix had already decided to produce its own transformers.[22] It would therefore be logical for it to produce its own cathode-ray tubes, though it would be anything but easy to design and manufacture tubes significantly better than those the two suppliers offered. In the end, the near impossibility of the task was the main reason to undertake the project and to stick with it from early 1951 until 1954. During those years the project team encountered numerous dead ends, frustrations, and obstacles, only to emerge with a tube far superior to anything then on the market.[23] In some ways, the venture had been foolhardy, for the cathode-ray tube was the oscilloscope's critical component and failure would have been monetarily, emotionally, and intellectually devastating. As a success, however, it was a triumph for Tektronix culture and a sign of the company's emerging self-confidence.

Tektronix based its reputation on its product's performance, which in turn relied on quality.[24] A malfunctioning part could damage the company's reputation; hence the need to control the quality of the components by manufacturing rather than buying them. This obsession with quality is attested by the practice of having each assembler initial his or her work inside the oscilloscopes. It also lies behind what is probably the company's most famous story.[25]

In it, Vollum, by then president of his own company in the late 1940s, was phoned by a woman who complained that her vacuum cleaner, which he had repaired years before, had again broken down. Without protest, Vollum went to her home and fixed the machine—free of charge, of course. The anecdote is instructive, for though the event may not have occurred, it reflects the essence of Vollum's and his company's commitment.

If there was one thing that symbolized the coalescence of a distinctive Tektronix culture in the company's early years, it was the advent of profit sharing. The initial success of the Series 511 found the company unprepared for the avalanche of orders that poured in, causing production to lag chronically behind demand. By 1949, Tektronix strained to fill back orders that were between six and twelve months old.[26] In an effort to increase output, management instituted a production bonus, based on the monthly number of instruments shipped.[27]

When Tektronix brought out other model types, however, the production bonus became a cumbersome accounting problem and an inadequate incentive to workers.[28] About a year after it was started, the bonus was therefore abandoned and profit sharing instituted as a more practical and comprehensive measure to improve production and reward efficiency.

Profit share, as the system came to be called, gave Tektronix employees a feeling of ownership in the company and therefore helped to curtail pilfering, an occasional problem.[29] Pride in ownership also helped ward off the threat of unionization and reinforced the egalitarian atmosphere and the sense of pride that have often been noted by outsiders. In its first year, profit share totaled 25 percent of pretax profits and the feeling of being a large but closely knit family asserted itself.[30]

Later the familial feeling came to have a troublesome effect on Tektronix. When, despite rigorous hiring policies, a firing loomed, there was little taste for it at any level of the company. Consequently, employees who under a tougher regime would have been dismissed quickly were allowed to linger at Tektronix long after their usefulness to the company was exhausted. After all, they were members of the family. Company historian Lee remarks that this propensity became a Tektronix tradition.[31]

Such a tradition could be tolerated in halcyon days, but in times of recession, such as those of the early and mid-eighties, it was an internal threat. Thus it is said that the company laid off two thousand workers in 1986 not because of sudden diminished production or low sales, but because those employees had long since become superfluous. Yet even when a financial crisis forced managers to act, the company asked the workers themselves to quit or retire early.[32]

THE FIFTIES

The fifties proved crucial for Tektronix in several respects. In that decade, key events transformed the company from a struggling young Portland industrial concern into a Washington County giant, a world leader in high technology, and a major enterprise with a distinctive personality.

By 1951 Tektronix's net sales exceeded $4 million and the payroll encompassed over three hundred employees.[33] Already, there was general agreement among the members of Tek's board that the Hawthorne plant had become too cramped and that a move to a larger space should be considered.[34] In keeping with company tradition, the workers were consulted about their preferences. As there were relatively few prime yet inexpensive industrial sites in Portland's core area, the possibilities were limited to the city's eastern suburbs around Gresham or those in the west near Beaverton. The employees chose the west, perhaps in recognition of the pleasant and diverse housing developments that had begun to appear there since the war. On 26 April 1950, therefore, the company acquired four acres of land at the intersection of Barnes Road and the Sunset Highway just beyond the western slope of Portland's West Hills. The sale price was $9,000.[35] By fall 1951 Tektronix was doing business in a twenty-thousand-square-foot, $200,000 Sunset plant there.

Soon even these expanded quarters proved inadequate. By 1954 nearly seven hundred people worked at Tektronix and net sales had reached $7.3 million.[36] In three years, the company had nearly doubled its size and sales. The demand and uses for oscilloscopes had grown enormously, spurred on in large part by the military's demands during the Korean War and by the post war industrial and commercial booms affecting the nation's economy.

In response to these conditions, the company had subjected its Sunset plant to constant modification and expansion, but the measures were never adequate. Consequently, when 313 acres of land west of the Beaverton city limits became available in 1956, the company bought them, for $396,000.[37] By the summer of 1957, the first building, a warehouse, had been constructed on the new grounds. Eventually a series of buildings in a "campus" environment, featuring large expanses of open space and wide avenues, emerged as the main setting for Tektronix's activities. However, the firm's physical growth was by no means at an end.

Tektronix's rapid expansion was not without costs. When 1954 began, the only original shareholders still associated with the company were Murdock and Vollum. In 1948 Glenn McDowell had exchanged his stock for Hawthorne Electronics, the retail arm of the firm, and had left to run it. Then, in 1953, Miles Tippery and Milt Bave left, together with Logan Belleville, an engineer who had held a significant interest in the company since 1948. Although their reasons for severing ties with the company varied, all three were known to feel that Tektronix had become too large.[38]

With Vollum and Murdock in control, the company remained in good hands. In addition, as it expanded, it had attracted a number of people who together formed an impressive managerial, marketing, and engineering nucleus. These included Bob Davis, an engineer whose first contribution was to organize the company's production line and who eventually became its first executive vice-president; Warren "Dal" Dallas, its first sales manager; and Dick Ropiequet, Derrol Pennington, Dick Rhiger, John Kobbe, Cliff Moulton, and Bill Polits, who filled various roles in research and development.

The value of these members of the infant company cannot be overemphasized. Together, they represented a reservoir of talent, imagination, and foresight for the two partners to draw on. Individually, they provided keys to the future. Davis's insistence on rational and swift production, for example, was a small but essential move forward. Kobbe's unblanking device put the company a step ahead of the competition. Over the years, the other engineers also made critical contributions. It was left to Dallas, however, to make a real revolution in electronics.

In the postwar years the electronics industry's marketing and sales efforts were universally characterized by a reliance on independent distributors, who represented a variety of products to customers. Some members of these organizations' sales forces were capable technicians, but for the most part they simply sold the product. Once an instrument was sold, they had finished with it. However, as new electronics products—particularly television—emerged in the late forties and early fifties, a field force of technicians who could service the sophisticated equipment of the distributors was increasingly needed. Such a group could quickly respond to technical problems, educate customers to new generations of electronics equipment, and maintain a high level of satisfaction with the product.[39]

Dallas understood this and realized that, if the sales force were combined with the service force, the market for the Tektronix oscilloscope would dramatically expand.[40] He therefore proposed that the company should create a network of "field engineers," who would be both salespersons and service technicians.[41]

In early 1951, after Dallas had expended considerable time and effort on creating a distributor network, Tektronix opened its first field office in New York City.[42] By mid-decade the company had an excellent and extensive field force in place throughout the nation. The Tektronix model became the standard of the electronics industry, and most other major firms eventually followed its lead.[43] As Dallas had found, the field office was crucial in the expansion of sales.

It was also Dallas who, with the support of Davis, expanded the company's horizons beyond the borders of the United States to the British Isles and Europe. In the mid-fifties, with the European recovery from war gathering momentum and industrial activity resuming prewar levels, Tektronix found a booming market for its oscilloscopes in Great Britain and on the continent. The formation of the Common Market in 1958, however, and the subsequent response of the British, who had been excluded from its ranks, threatened to disrupt the company's burgeoning European business.[44] The upshot of British exclusion from the European Economic Community had been a round of protectionism that had resulted in Britain's establishment of "Commonwealth Preference," which allowed goods made largely of materials from Commonwealth countries or by mainly Commonwealth labor to enter Great Britain duty free. All other merchandise would be subject to a heavy tariff. Although aimed primarily at Common Market goods, the rule had affected Tektronix as well. Great Britain was the company's foremost overseas market, and one of its major competitors, Cossor, was a British firm. Faced with Commonwealth Preference, Tektronix could hardly hope to compete against Cossor.[45]

Although overseas sales were only an afterthought to both Murdock and Vollum, Dallas gave the problem serious attention. He recognized that the company had a single option if it wished to continue selling in Europe: it must remain competitive by opening a branch plant in Britain to make oscilloscopes that would qualify under Commonwealth Preference.

In his effort to persuade Vollum and Murdock of the need for a plant, Dallas enlisted the help of Davis. Together they succeeded, and in 1957 Dallas hired an American expatriate who was experienced in electronics, Al Hannman, to find a suitable site for the new venture.[46] For a number of reasons, Hannman pinpointed the Isle of Guernsey as the most appropriate site. A member of the Commonwealth, Guernsey had a duty-free port and—because of an agricultural depression—a stable, industrious, yet largely unemployed and therefore inexpensive labor pool.[47] With goods able to enter both Common Market and English ports duty free and with cheap but hardworking labor available, it was the perfect spot for the new venture.

By 1960 more than seventy people were employed in Tektronix's Guernsey plant, which had outperformed every projection. Much of this was due to the diligence of the work force, the thoroughness and soundness of Davis's planning, and the excellence of the management team imported from Oregon.[48] A linchpin of Guernsey management was Earl Wantland, a 25-year-old whom Davis had plucked from the Beaverton Test Department to be production manager on the island. Wantland's intelligence, energy, and leadership potential were genuine, as his record on Guernsey was to indicate. Those qualities were to prove invaluable in the eighties, when Wantland, as president of the entire company, was faced with some of Tektronix's most intractable problems.

A few years before the Guernsey plant opened, Tektronix would probably have been unable to mount such an initiative, even if revenues and markets had justified the idea. But by 1958 the company was no longer inexperienced and mature leadership had emerged in all its key areas. A failure in an endeavor such as Guernsey might have been a setback, but would not have doomed the company.

Tektronix's self-confidence in the mid- to late-fifties cannot be attributed only to the good health of the balance sheet. It was due equally to the leadership of Murdock and Vollum. Although in many respects Murdock had begun to distance himself from the company's day-to-day operations early in the decade, he remained fascinated by the problems of employee relations that became daily more and more critical to the organization's success. His interest in psychology first led him to engage doctors from the Menninger Foundation to help structure employee relations, particularly where the supervisor's role was of concern, and eventually to retain a full-time staff psychologist. Between the Menninger consultants and the company psychologist, John Wallen, Tektronix grew as humanely and unstressfully as possible despite the pace of expansion during the decade. Murdock was careful to nurture the highly prized egalitarianism and familiarity of the company's early days, while making sure that supervisory personnel were strong and creative leaders.[49]

As Murdock participated in the firm selectively, Vollum immersed himself in it more deeply and pervasively. The company virtually centered on him, and it increasingly took on his traits. His humanity and simplicity gave Tektronix its character, even as it grew to the anonymous proportions of a large institution. His integrity, thoroughness, and perfectionism were reflected in the reliability of each oscilloscope that came off the production line. His interest in technical advancement led from the company's first rudimentary 501 to the 530 Series of the mid-fifties. The latter were highly sophisticated instruments involving radically new modular concepts in design that provided a versatility previously unheard of in oscilloscopes. Typically, Vollum was actively involved in the design and development of these "plug-in" instruments, as they were called.

As the decade ended, the still-young company had continued to expand (in 1959 the payroll was up to 2,565 workers), but its vigor and optimism were unimpaired.[50] Now cosmopolitan where once it had been merely local, it was led by able and assured men whose vistas had at first encompassed only modest, uncertain possibilities.

THE SIXTIES AND SEVENTIES

In the sixties and seventies, Tektronix took some remarkable steps that would enlarge its already impressive share of the world oscilloscope market.

A priority was to expand its European operations. The original hope that the Guernsey plant would supply all of Europe had proved unrealistic. It was thus deemed necessary to create an international network of plants and marketing operations that would increase production and maximize profits.[51]

With expert advice to guide them, Tektronix leaders embarked on a plan to create a subsidiary rather than a branch operation that, for tax purposes, would be under the control of a holding company and linked trading company. In March 1961, an entity known as Tekintag (Tektronix International, A.G.) was formed, a holding and trading structure, headquartered in Switzerland.[52] The first offspring of Tekintag was a plant at Heerenveen in the Friesland region of The Netherlands. The plant began production in spring 1962 under the direction of Wantland.

Over the next two decades, this venture evolved into a highly successful operation that was strengthened through the acquisition of independent distributors in key countries such as France, Italy, and the United Kingdom. By the end of the seventies, the company acquired a respected sales and service network throughout Europe.

Tektronix was not only making its presence felt in Europe. Through a domestic distributor, the company had been marketing its oscilloscopes in Japan. However, because of the vagaries of that country's import laws, its share of the Japanese market was modest. As early as 1960, Tektronix's international accounting consultants had confirmed its leaders' belief that to exploit the potential of Japanese demand, the company would have to enter into a partnership with a

Japanese firm.[53] Several major Japanese companies had already proposed such a relationship to Tektronix. But they had pirated Tektronix's designs for their own oscilloscopes (which Vollum considered unethical), and none had offered the company more than a junior status in the relationship. In 1963, in order to discuss the proposals of Toshiba and Hitachi, Vollum traveled to Japan. While there, he explored additional possibilities and found that the Sony Corporation had some interest in a partnership.[54]

By 1965 Sony was a respected name in international business circles, associated with innovation, quality, and shrewd marketing. In March, after extensive talks between the two companies and the development of a friendship between Vollum and Akio Morita, Sony's founder and chairman, the two struck a deal. The result was the Sony-Tektronix Corporation, constituted under the joint presidency of Vollum and Morita and the joint managership of Takashi Kumakura of Sony and Howard Mikesell of Tektronix.[55] The deal also involved a fifty-fifty division of profits, a principle for which Morita was forced to fight with his own government.[56] The entire arrangement was an unheard-of triumph over the Japanese economic blueprint and the sensibilities of its planners, but it has persisted into the eighties as a perfectly functioning model of the joint venture.

During the worldwide expansion of Tektronix, events occurred in Beaverton and elsewhere in the United States that would have an equal impact on the company. Among them was the company's reorganization, catalyzed by a form of one-man rule that had become the style at Tektronix under Bob Davis, the executive vice-president and chief operating officer. An able administrator with a strong personality and an apparent need to be involved in everything that happened at the firm, Davis had become the critical element in every initiative taken at Tektronix.[57] A serious problem had resulted from this propensity: the management team that had emerged at the end of the fifties had been unable to perform efficiently. That, in turn, had hurt the company's performance.

Vollum and Murdock had originally been satisfied that Davis had everything in hand. However, once he realized what was happening, Vollum consulted the management committee and then relieved Davis of his duties. In June 1962 he reemerged as the company's chief operating officer.

Murdock, on the other hand, continued to withdraw from Tektronix. Perhaps alienated by a lack of understanding and interest in the nuances of electronics research, as well as by the size and complexity of the company he had created, he seemed to feel superfluous. Although he still served as chairman, he gave most of his time and attention to other diversions, one of which cost him his life. On 16 May 1971, he drowned in the Columbia River near Maryhill, Washington, when the seaplane he was piloting crashed during takeoff.[58]

At about the time of the reorganization, however, Murdock and Vollum had other fateful decisions to make, among the most significant of which was the one to go public. On 11 September 1963 the firm was transformed from a closely held corporation to a public one, when stock was offered over the counter for the first time. On 10 January 1964 the symbol "TEK" initially appeared on the "Big Board" of the New York Stock Exchange at 23 1/8.[59] Tektronix was the first company incorporated in Oregon to gain this distinction.[60] Continued growth also allowed the company to enter the elite circle of *Fortune* 500 companies in May 1976.[61]

Tektronix had made another pivotal decision in 1961, when it initiated litigation against the federal government for patent infringement. As early as 1958, the military had contracted with several firms to produce oscilloscopes that were much less expensive than those of Tektronix. All those concerned were unwilling to reimburse Tek for the use of its patent rights. Only after eighteen years was the suit settled, but its successful outcome for the company was beneficial to the entire industry.[62] The principle Tektronix established was simple: the federal government would have to pay for the patents it wanted to use.

In 1964 Tektronix acquired its first company, a Brooklyn, New York, firm called Pentrix, which made spectrum analyzers.[63] The Pentrix product had been designed to plug into the Tektronix oscilloscope and therefore represented a natural acquisition for the Beaverton company. A year later Tektronix presented the Series 491 oscilloscope, a general-purpose, portable spectrum analyzer.

If the Pentrix acquisition was a shortcut to dominance of a relatively new but associated field, it was also a tentative step toward diversification. Yet diversification was not really a goal at the time—an oversight for which the company would later suffer.

By the mid-sixties Tektronix was lingering in a dangerous phase, during which complacency about its basic product had taken hold. Despite the sterling reputation of its oscilloscopes, increased competition and a creeping obsolescence loomed as threats to the company's market dominance. In the early months of 1964, Hewlett-Packard's introduction of a new oscilloscope with capabilities greater than anything Tektronix offered was particularly alarming to Vollum.[64] Although his engineers remained unconcerned, he ordered that research begin on the next generation of Tektronix oscilloscopes.[65]

The result of his intervention in the research schedule was the frenzied development of the Series 7000. After five years characterized by the founder's continued pressure on the engineering teams assigned to the project, the new oscilloscope emerged in 1969, once again as the industry's standard. The two models, the 7504 and 7704, could both take four modular plug-in units. In addition, there were thirteen plug-in accessories, and a design for lighted push-button switches that has never been duplicated. Automatic triggering, functional color coding, and dual-trace switching between channels in the oscilloscope rather than the plug-ins were other innovative features.[66] In short, the instrument more than met Vollum's requirements: it set the direction for the future.

In 1969 Tektronix also introduced a product that represented a significant step toward diversification: the T4002 Graphics Computer Terminal. The product was the brainchild of C. Norman Winningstad, an engineer who had joined Tektronix in 1958. Bringing together researchers who had worked on two disparate projects—semiconductors and storage tubes—he began the Electron Physics Research Group, which was dedicated to tapping the emergent market for display devices. The result was not only the T4002, but release from dependence on the oscilloscope. Tektronix had become more than a one-product company.

For Winningstad, however, there was little to savor in the victory. In January 1969, when Wantland was named executive vice-president and a major reorganization of the company occurred, a reassessment of various initiatives was made. Because of high costs relative to sales projections and actual sales in display devices, Winningstad was passed over when the Information Display Products Group was formed, and his group was subsumed.[67] He left the company and started Floating Point Systems, often called the most successful Tektronix spin-off.

Diversification, innovation, reorganization: they were exhilarating experiences in many ways, but they left the company unprepared for the crises of the early seventies. The first was the death of Murdock. The second was the revelation of a nearly 12 percent slide in net sales, which caught Tektronix with its largest inventory ever.[68] The third was Vollum's heart attack, which occurred at about the same time. Vollum returned to work as the newly elected chairman in September 1971, but relinquished the presidency to Wantland the next year. The company's financial health was less resilient. It had begun to decline as early as 1969 and would not return to normal until 1974, partly from stagnation in the domestic economy, caused by the "guns-and-butter" policies of the Johnson and Nixon administrations, and partly from a slump in the global economy, because of the OPEC oil embargo of 1973.

The economic deterioration eventually catalyzed a greater reorganization than had been intended in early 1969, when Wantland had been made sole executive vice-president. The initial reorganization had been confined to the management level, but by 1971 the entire company was being restructured into seven product groups, all supported by centralized services consisting of manufacturing, engineering, administration, and finance.[69]

In August 1972 the reorganization was carried one step further with the creation of the Tektronix Laboratory, which was charged with the research and development of integrated circuits, solid-state devices, and cathode-ray tubes.[70] The subtext of the notice announcing the laboratory's formation stated that it was to carry on wider ranging, more innovative research than Tektronix had ever undertaken.

Even before then, some progress had been made, spurred by a growing sense that new products and technological breakthroughs were essential to survival. In 1971, for example, Tektronix unveiled the Series 7904, the fastest general-purpose oscilloscope. It also introduced the S103, a low-frequency oscilloscope featuring three plug-ins and a choice of four kinds of cathode-ray tubes. The establishment of the laboratory stimulated the development, throughout the decade, of an array of new, daring, and imaginative products that the company might not otherwise have developed.[71]

Perhaps the chief beneficiary of the move toward diversification was the Information Display Group. Under the direction of Larry Mayhew it grew from a minute part of Tektronix's business to an entity that brought in $63.6 million in sales, representing 17 percent of the company's total sales in 1976.[72] In 1975 it had become so big that it moved into a new plant in Wilsonville, south of Portland. It was the company's first domestic manufacturing operation away from the Beaverton campus.

Information Display was a symbol of the growth that diversification and innovation could bring once the shackles came off the global economy. By the end of the decade, four other new divisions sprang up: Communications, Systems, Laboratory Instruments, and Tests and Measurements. Yet the last half of the seventies and the first half of the eighties were to be marked by the difficulties of coping with the effects of Tektronix's own success.

THE EIGHTIES

On Monday, 3 February 1986, Howard Vollum was rushed to St. Vincent Hospital after suffering a massive stroke during a dinner with friends.[73] There, within view of the first Washington County site of Tektronix, he died.

At the hour of its key founder's death, Tektronix was a successful and honored company. However, viewed beside the energy, enthusiasm, daring, and innovation of the new high-technology entrepreneurs, the company seemed lethargic, conservative, encumbered by institutional inertia, uninformed by its past clarity of scientific vision, and bereft of a shaping dynamic.

This interpretation was strengthened a few months after Vollum's death, when readers of the 22 May *Oregonian* were greeted by a front-page headline of shocking singularity: "Slumping Tektronix to Cut Force by 2,000 Jobs."[74] The accompanying article noted a 7 percent decline in sales and a 48 percent earnings decline for the year.[75]

The travails of the mid-eighties, however, seemed at the time to be balanced by a spate of initiatives that had only recently been taken or had not quite begun to pay dividends. Among these was divisionalization itself, which promised to affirm the diversification necessary to the company's prosperity. There was also the Tektronix Development Company, which had nurtured such firms as TriQuint Semiconductor, Planar, and Anthro Corporation.[76] And there were expanded foreign operations in such diverse countries as Brazil, India, and China.[77]

Some of the most significant steps of the eighties were concealed from public view by their unspectacular nature. They included techniques known as Manufacturing Resources Planning and Just in Time Manufacturing, which sliced inventory by $100 million from 1981 to 1985, turned it over more than six times in 1985 compared with 3.6 times in 1981, and nearly doubled employee productivity measured by sales in a comparison of the two years.[78]

In 1985, its fortieth year, then, Tektronix found itself struggling toward a new stage of life, with a broadened identity and a leaner image. Without Vollum, the founder and personification of the values of its past, progress would be difficult. The heirs to his mantle were leading vigorously, if with initially mixed results. As the legatees of two young men's modest dream that was at the center of a state's economy, the fountainhead of an entire regional industry, and a legend of American business, they faced an enormous responsibility.

4

Floating Point Systems

FLOATING POINT SYSTEMS REPRESENTS the second generation of Washington County high-technology firms after the pioneers, Tektronix and ESI. It is significant as the first successful "spin-off" company formed by employees who left their established firm; as the third of the county's corporations to be listed on the New York Stock Exchange; and as the first Washington County company that prospered initially by manufacturing products not used for tests and measurements.

Floating Point was founded by C. Norman Winningstad, who was born and raised in Berkeley, California. As a boy, he wanted to be like both his grandfathers and become an engineer and he was fascinated with building model airplanes. Later at the University of California at Berkeley, these two early interests came together in a mechanical engineering major, which in those days would have been an appropriate starting point for a specialization in aeronautical engineering.[1]

Like Douglas Strain, however, Winningstad was caught up in the Second World War before he could finish college, and—again like Strain—his wartime experience was important to his future in high technology. When he joined the Navy, he became an electronic technician's mate, which spurred his interest in electronics; and when he returned to college following the war, he switched his field to electrical engineering. He graduated from Berkeley in 1948 with a specialty in vacuum tubes. After graduation, he worked for two years on an experimental television project for California Television and for another eight at the Lawrence Radiation Laboratory in Berkeley. In 1959 he came to Oregon to work for Tektronix.[2]

At Washington County's first high-technology firm, Winningstad rose rapidly as an engineer and as an administrator, publishing several papers, obtaining six patents, and developing two lines of more than eighteen products, with sales of over $25 million. Although he was successful in his work, in 1970 he began to look for new challenges and opportunities. He resigned from Tektronix in that year to enter the MBA program at Portland State University and in October joined three partners—Tom Prince, Frank Bouton, and Robert Carter—to found Floating Point Systems in Beaverton, then a town of 18,577.[3]

The original idea for the company was that of Carter, a salesman for Data General Corporation. He persuaded Bouton and Prince, who were working at Tektronix, of the validity of the plan. Winningstad was the last to join. Tektronix had no objection to the new venture, which would not compete with its product line.[4]

No one remembers who chose the name of the new company, but the reason for doing so is clear. The first computers and calculators did fixed-point arithmetic—that is, users fixed the decimal point for their calculations before doing them. For example, a calculator user dealing with whole numbers could set the decimal point two places from the right for dollar arithmetic or four places from the right for calculations requiring greater accuracy. The method was "fixed" because the decimal point never moved from the position the user selected. It was a satisfactory system for the world of business, but inadequate for scientists and engineers, who dealt with enormously varying sizes of numbers and who could not usually predict where the decimal point should be. What these professionals needed was an arithmetic that allowed the decimal point to "float."

At Floating Point the partners began by manufacturing a hardware attachment that improved the performance of the minicomputers that were just appearing on the market. With this attachment, computers could do floating-point arithmetic twenty times faster than they could without it.

Despite the significance of its product, the company began in a somewhat modest and precarious manner. The four principals raised their first capital from their own assets (with Winningstad contributing $300,000, including his helicopter and airplane), hired five employees, and went to work in a converted Quonset hut in Beaverton. With the characteristic speed of successful high-technology companies, however, they shipped their first product eight months after incorporating, in June 1971. By 1972 the company had made two geographical moves and realized its first profit. Nevertheless, all three of the first facilities were unprepossessing. (Winningstad paddled a canoe in the flooded parking lot at one of them, which was so seedy that a customer's engineer told them not to invite any of his headquarters people there.)[5]

THE VENTURE INTO ARRAY PROCESSORS

Two years later Floating Point Systems changed direction sharply. It had reached a turning point because it lacked the capital to continue expansion of the original product line and because the minicomputer companies were beginning to bring out their own floating-point hardware. For lack of capital, Winningstad's first three associates (who, unlike him, were not independently wealthy) dropped out, while Winningstad decided to embark alone on a new product, the array processor, although he had to mortgage his home to do so.

The company first ventured into this field when Control Data Corporation asked to have floating-point hardware designed for one of its models and later requested that Floating Point Systems design for it a special piece of equipment, the array processor. Such a processor had been invented for NASA by Glen Culler, a scientist at California's Ames Research Laboratory. It represented one of the most interesting developments of the electronics age because it addressed in a new way its very symbol, the computer.[6] While floating-point hardware processes two numbers at a time and returns one answer, an array processor performs thousands of calculations on numbers at the same time and returns thousands of answers. The numbers come to the computer in the form of an array—an orderly arrangement—and the answers are also returned in that form.[7]

Winningstad realized that the array processor allowed business computers to be inexpensively converted for use by scientists needing to arrange data and to perform sophisticated mathematical calculations. Without the array processor, the financial implications of providing computers for both business and science were overwhelming for most corporations and universities, which might ultimately be forced to choose between the two interests. Array processors solved the dilemma by converting minicomputers into scientific computers without the enormous cost associated with "handmade" machines.

The catalyst for Floating Point's entry into array processing was the oil cutback in 1973, which touched off an extensive search for new oil reserves in the United States. Floating Point's vice president of engineering, George O'Leary, knew that an array processor was what the oil companies needed, but he also knew that it would take too long to redesign his Control Data Corporation machine and produce it for a new market. The problem was resolved when Winningstad contacted Culler, an old acquaintance, and found that he had designed an array processor that could be easily modified for the oil companies' use.[8]

O'Leary and his colleague Alan Charlesworth, an applications engineer, began to modify Culler's array processor in 1974, the year the company moved into a warehouse in Tigard. Within nine months Floating Point had designed, produced, and sold its first array processor, the AP-120B. The first customer was a seismic exploration service company in Houston which took delivery in July 1975. Floating Point had had to move quickly because when it entered the market, it already had five significant competitors, from giants like IBM to small start-ups. However, it was soon selling its AP-120B model to such firms as Chevron, AMOCO, Geosource, and Western Geophysical, and its sales were to increase in value fifty times from 1975 to 1980.

Its success with array processors enabled the company to close its floating-point hardware business in 1975 and move in the next year to a plant in Beaverton, which it styled its "world headquarters," in parody of IBM's building in Armonk, New York.[9] Under Charlesworth's direction, the company also developed computer software that outperformed that of the competition. By the end of 1976 Floating Point had a popular product, a strong marketing network, and a sound financial basis.

In the next few years other uses were found for the array processors, the best known being in x-ray tomography. Original x-ray techniques (shadowgraphs) were unsatisfactory in examining brain tumors and cancers (such as pancreatic cancer) that involved the need to achieve contrast. Tomography replaced x-ray film with more accurate computer analysis and display of the x-ray data (the CAT-scanning technique). In addition to its uses in medicine, the array processor had applications for the armed forces, astronomers, cartographers, and meteorologists, among others. The Floating Point array processor was also useful in simulation processing. Examples from a wide variety of modeling applications include aircraft flight simulators, electronic-circuit analysis, and NASA space-shuttle simulation. In 1978 Floating Point introduced a new model, the AP-190L, which was designed to link up with large mainframe computers such as those manufactured by IBM and Sperry Rand.[10]

GOING PUBLIC: GOOD NEWS AND BAD

In 1978 the company was flourishing. It broke ground for new headquarters in Beaverton, achieved record product sales in the fourth quarter, and tapped a new source of capital. Its first public offering of common stock, 750,000 shares (588,284 by the company and 161,716 by certain shareholders), was made in September and was received enthusiastically, providing it with approximately $9.5 million in additional capital. The significance of the public offering was not only to increase working capital 6.7 times but to raise stockholders' equity, eliminate short-term debt, and improve the total debt-to-equity ratio. The offering made it possible to enlarge the plant facilities to meet the corporation's burgeoning sales, which had quadrupled not only in 1976, but again in 1977 because of a contract with General Electric and increased the company's net worth so that the land for expansion could be leased and the new building begun.[11]

In 1979 Floating Point Systems entered a somewhat troubled period, although new products were being introduced and the new headquarters opened. Trouble began at the first of the year, when seventy-nine employees were laid off because expected sales in the x-ray tomography market had not materialized. In May, the Enterprise Wallpaper Manufacturing Company filed a class-action suit against Floating Point for alleged damages sustained during the initial offering of public stock. When the stock was issued on 28 September 1978, its price was $18 a share. Within a few days, the price had risen to $26, then had fallen to $12 when the company reported an anticipated loss for the first quarter of 1979. The lawsuit charged violations of federal securities laws through failure to disclose adequate information in the stock offering prospectus and the period following the initial sale. In November 1980 the suit was settled out of court with the agreement that warrants enabling holders to buy stock at 80 percent of market value would be distributed to members of the class who could properly document their losses. Floating Point also agreed to pay $200,000 in plaintiff's attorney fees and litigation costs. Although the settlement was expensive, it removed a cloud of uncertainty from the company's future.[12]

THE EARLY EIGHTIES: NEW PRODUCTS AND GROWTH

In other respects 1980 was a successful year. The company announced a new product, the FPS-164 Attached Processor, which had greater precision and memory than did earlier Floating Point products. Development of the FPS-164 was an example of management foresight at the company.

By the end of the seventies, recovery from the oil crisis and a consequent reduction in petroleum exploration made it clear that new products had to be developed to supplement the oil market. Moreover, signal-processing markets, such as sonar and x-ray tomography, were beginning to stabilize. The time had come to develop other computers besides array processors. Enter the FPS-164.

A scientific computer (sometimes called a minisupercomputer) the FPS-164 was designed for what was anticipated to be an unlimited simulation market. (In fact, this market accounted for less than 10 percent of Floating Point's sales in 1981.)[13] By 1980, simulation, often called computer-aided design (CAD) or computer-aided engineering (CAE), was a recognized field, but most CAD had been accomplished on large mainframe computers costing between $4 million and $10 million. By contrast, the FPS-164 and its host computer cost only $600,000. If the computer proved popular, Floating Point's management reasoned, many firms would find computer design faster, cheaper, and often more accurate than scaled-down physical models.

Since CAD can be used in aircraft and automobile design, it can replace the expensive wind tunnels and scale models traditionally used to test airflow. (Besides being costly, scale models do not always behave like their full-scale counterparts. Moreover, while changing the shape of a wing or a fuselage may take days of work by several people, an engineer at a computer terminal can make immediate design changes on a full-size mathematical model without reworking a physical one.) CAD is also indispensable in designing miniature electronic circuits and in analyzing a structure's ability to withstand physical changes over a range of loads and with changing loads over a period of time.[14]

In addition to product changes, marketing innovations also distinguished the year 1980. In February, Floating Point came to an agreement with Digital Equipment Corporation, the world's largest manufacturer of minicomputers, to obtain that company's actual and potential customer base in return for giving Digital the advantage of offering array processors in scientific markets. Because Digital and IBM predominated in the scientific-computer area, the agreement helped convince customers who were unsure about Floating Point that it was a sound corporation. An agreement with IBM further enhanced the company's credibility by providing that representatives of either company could set up joint sales sessions with computer users who might need Floating Point's array processors. (It was the only agreement that IBM had made with an array-processing company.) Later, other agreements followed with Sperry and other companies. Then, in March, Floating Point made an agreement with a French computer company—CII Honeywell Bull, a government firm—giving the company entry to government sales of its computer in France.[15]

These changes marked Floating Point not only as a successful corporation, but as a rapidly growing one. Its growth was so rapid that Winningstad retired from the presidency in 1981 to devote time to his duties as chairman of the board and chief executive officer, which he had held simultaneously with the presidency. The move left him free to concentrate on contacts with bankers and financial analysts and on investor and government relations. Contributing to this decision was his earlier realization that at neither Tektronix nor Floating Point had he managed operations with more than $20 million in sales. As Floating Point moved toward the $40 million mark, he was afraid that he could not "learn fast enough to keep up with the company."

As early as 1977, he had begun to look for a successor—someone who had run a $100 million organization that had grown by a factor of at least two during his tenure and had been a line manager in two of three areas: manufacturing, engineering, and marketing. The man who met these requirements was Lloyd D. Turner, vice president of engineering for Burroughs Corporation, who had the advantage of also having worked for a small computer company.[16]

Turner assumed his duties as president and chief operating officer on 1 June 1981. He was in his new position when the FPS-164 went into production, when the company was listed on the New York Stock Exchange on 6 October, and when it made its second public stock offering ($24 million, which more than doubled stockholders' equity) to make possible the development of a sixty-acre tract west of its new building.

The year 1981 also marked Floating Point's decision to build a manufacturing plant in Ireland. Underlying the move was the fact that the company needed to build a plant closer to its European customers and to establish a physical, as well as a psychological, presence in Europe. While Floating Point was looking around, Winningstad was visited by an American representative of the Republic of Ireland's Industrial Development Agency, who persuaded him of the advantages the Government of Ireland could offer—for example, building grants (which offset the high costs of construction) and favorable tax provisions—as well as the wisdom of accepting them before Ireland was required by the European Common Market to end subsidies. These incentives, the common language, and the Irish worker's reputation for diligence carried the day, and a site in Swords, near Dublin, was soon selected. George O'Leary, as vice president of European Operations, moved to Ireland to oversee the project.[17]

TESTING COMPANY STRENGTH

After Floating Point's first fourteen years, the company overall was a testimony to the adaptability and viability of the high-technology industry. It had passed the formative period and the more difficult task of growth thanks to its management's success in devising new products for developing markets. All these qualities were to be tested in the great upheaval of 1984 caused by the loss of a lucrative General Electric contract to Star Technologies in February of that year.

Star was founded in August 1981 by sixteen persons, ten of whom were former employees of Floating Point Systems. They were led by Jon Salquist, who had been vice president of marketing and international operations, who became president of the new company. Star manufactured array processors, using a different technology from Floating Point's. In October 1982 the new company obtained its first order, a $10 million contract with Control Data Corporation. The firm gained some national publicity and then in February 1984 obtained a $44 million contract to supply CAT-scanning equipment to General Electric Medical Systems, beginning in 1985. When Floating Point announced the loss of this contract, which constituted 44 percent of the company's 1983 sales, not only Washington County but the stock market was staggered. At the opening of the New York Stock Exchange the day after the announcement, trading of the company's stock was delayed for three hours. When the exchange closed for the day, Floating Point's stock had fallen 19 percent from the previous day, leaving the company as the biggest loser on the exchange for the week.[18]

The company's innate strength was displayed in its rebound from this disaster. In March 1985 it sold five of its 264 scientific computers (for $650,000 each) to Cornell University for its federally funded supercomputer system. Cornell was to link the 264s with IBM mainframe computers to achieve high-performance systems.[19] This good news, though, was not as interesting to the electronics world as were rumors about the company's projected entry into the world of supercomputers with a brand-new machine (see chapter 7).

In a kind of cyclical pattern, however, bad news followed. On 22 July 1986 Floating Point's stock fell twelve points on the New York Stock Exchange, a loss of 37 percent of its value. The drop came after the company announced that several anticipated large orders had not materialized. Later that week two out-of-state shareholders sued the company, claiming that, beginning in October 1985, Floating Point had begun to issue overly optimistic reports about its prospects. A week later a second suit followed. The fall in stock prices also raised questions about the sale of large blocks of Floating Point stock by company officials (Winningstad, Turner, and three vice-presidents) two months before the bad news was released. All subsequently denied that "inside information" had caused them to make the sales in the knowledge that stock prices would later drop. On balance, these unsettling developments did little to tarnish the company's rise to prominence as a model of what the small entrepreneur could do in the high-technology world of Washington County.[20]

5

New Ventures

IN THE LATE SEVENTIES and the eighties, after the success of the "big three" electronics firms—Tektronix, ESI, and Floating Point Systems—was apparent, they were joined by the first out-of-state corporation—Intel—and a host of new high-technology ventures. The reasons for these developments are numerous.

Availability of land, adequate supplies of good water, and an intelligent and hardworking labor force continued to be attractions. (According to a later American Electronics Association study, Oregon had the lowest employee turnover rate among electronics engineers in the nation.)[1] Cooperation among real-estate developers, development associations (such as the Sunset Corridor Association and the Tualatin Valley Economic Development Corporation), and local governments ensured a supply of good, properly prepared land.

Employees could be retained and new ones attracted to Washington County because of reasonable housing costs (far lower than in California), an excellent school district, and the state's natural beauty. For many, Oregon seemed a paradise, especially when contrasted with California or Massachusetts.

As we shall see in chapter 8, government itself encouraged high technology. After the gubernatorial years of Tom McCall (1967-75), when Oregon was perceived as being hostile to economic development, state officials proclaimed the state again "open for business." Particularly after the 1979 depression, the state began an attempt to alter its tax structure, complete its land-use regulations, and engage in other economic-development ventures. Pushed by the local council of the American Electronics Association, state government also resisted efforts by environmentalists and feminists (who were concerned about the welfare of the industry's many women workers) to pass legislation perceived to be injurious to the high-technology interest.

At the local level, city, county, and service districts were also responsive to the industry. It became easy to secure development permits, land-zoning changes, and other essential requirements. The county did what was needed to attract the new industry's payrolls. In turn, high technology successfully influenced the politics of the national government, its most notable success being the 1978 reduction of the capital-gains tax, which led to greater availability of venture capital. That capital was a major ingredient in the growth of the county's new start-up and spin-off firms, completing what commentators on high technology liked to call a "critical mass" of attractions—from the number of high-technology firms and engineers to educational opportunities and physical infrastructure. Above all, critical mass ensured the interchange of ideas necessary to economic progress.

INTEL

The first major high-technology firm in Washington County was a branch of one of the industry's giants, California's Intel. Its story begins with that of one of its founders, Robert Noyce.

In 1956 Noyce, who had grown up in Iowa and had a Ph.D. in electrical engineering from MIT, left an unsatisfying job in Philadelphia and flew to Palo Alto with his wife and two young children. There, he successfully applied for a job at Shockley Semiconductor Laboratory.[2]

The firm employed eleven other young scientists, including a chemist named Gordon Moore, who helped William Shockley in his research. Shockley—who was a co-winner of the Nobel Prize in physics for his invention of the transistor—was so idiosyncratic that eight of his assistants eventually found his quirks too wearing and decided to strike out on their own in the growing field of electronics. Among them were Noyce and Moore. In 1957, the group found financial backing for their enterprise from Fairchild Camera and Instrument Corporation of New York and set up Fairchild Semiconductor in a two-story warehouse in Mountain View. There, they began producing transistors and expanding electronics research in the quest for new and better products. Near them were such companies as General Electric, IBM, and Hewlett-Packard, and Stanford University, all of which were in the forefront of electronics research and development.

It was a propitious time to begin a transistor business. Both the general public and the industry were greatly interested in the transistor's small size and consequent ability to "shrink" products in which it was used. This created a large demand for the products, which were novelties, and consequently for the transistors themselves. Under the circumstances, Fairchild was eager to underwrite the Shockley defectors.

In 1959 the initial payoff on the research Fairchild had funded was announced. Although Jack Kilby of Texas Instruments invented the first solid-state, or integrated, circuit six months earlier, in July Fairchild announced that Noyce had also independently invented one. Noyce's device, made of silicon, worked better than Kilby's and became the standard of the industry.[3] The invention made Noyce relatively wealthy at the age of thirty-one.

By 1968 Fairchild had established itself as a leading company in semiconductors. However, for a variety of reasons, Noyce and Moore had again become dissatisfied with their situation. With Andrew Grove, a young chemical engineer on the Fairchild staff, they planned another venture. This time they decided to pool their own funds in order to be the major stockholders in the new company. With the aid of Arthur Rock, the San Francisco investment banker who had secured Fairchild's participation in the 1957 venture, they attracted additional capital and set up their new company, named Intel, in a small building in Mountain View.

The company immediately began research in data storage, or computer memory, at the time the least developed aspect of computer technology. Within two years the founders had perfected the 1103 memory chip—a silicon and polysilicon device smaller, faster, and far more capable of storing information than any of the old-fashioned ceramic cores or ringlets that were used for holding memory in the computers of the day. Whereas each core held one bit of information, the 1103 stored one thousand.

The 1103 was the world's first 1K dynamic random-access memory and constituted a watershed in the history of the integrated circuit: large amounts of information could now be stored on one chip.[4] By 1972, it had become the largest selling semiconductor memory in the world.[5] Compared with later generations of silicon chips, the 1103 was crude: hard to manufacture, difficult to test, temperamental, and slow to operate. Still, it was a huge advance over core memories, vastly multiplying the power of contemporary computers. Moreover, it secured the reputation of Intel as a serious contributor to technological advancement in electronics.

The name of the new company was made-up, because it sounded "sexy" to the founders.[6] There was nothing made up, however, about the growth catalyzed by the 1103. From first-year sales of less than $3,000 and a staff of 42, Intel's sales had grown to $66 million and the staff to 2,528 by 1973.[7]

While work proceeded on the 1103, the company received a contract to design a set of chips for a group of high-performance calculators. A young engineer named Ted Hoff was assigned to head the project. To Hoff, the original design concept for the chips that the calculators' manufacturer had submitted seemed needlessly cumbersome and uneconomical. He therefore attempted to streamline it drastically by simplifying the instructions in the chips and providing a central memory device. The outcome was a tiny (approximately 1/6-inch by 1/8-inch), general-purpose computer programmed to be a calculator, instead of an instrument that was merely a calculator.[8]

The new design used four chips instead of the five the original device called for. At its heart was a central processing unit (CPU) chip, which came to be called the microprocessor. Made up of twenty-three hundred metal oxide semiconductor (MOS) transistors, the first microprocessor (the 4004, as the series was called) rivaled the capacity of the original electronic computer, ENIAC.[9]

While the electronics industry hardly took notice of the microprocessor, Intel engineers recognized it as a real innovation—an entirely new kind of computer with a great commercial potential.[10] With its versatility and low cost, it could introduce intelligence into machines.[11] It could, and eventually did, change information gathering, communication, and work methods. In short, it was the linchpin of an electronics revolution—in the words of one electronics expert, perhaps "the most significant invention of the twentieth century."[12]

Between the success of the 1103 and the eventual acceptance of the microprocessor, Intel established itself as a leader in the computer industry. Its reputation and its business continued to grow. By 1974 the company had booked about $200 million worth of orders.[13] The founders began to think broadly about the terms under which Intel should expand. The company had moved from its original Mountain View location in 1970 when it had purchased twenty-six acres of land in the adjacent city of Santa Clara with the idea of building a larger facility. But the new, as yet unformed plans promised even greater growth.[14]

Moore, Noyce, and Grove began with several assumptions. For a number of reasons they knew they needed to move beyond the Santa Clara Valley. With success had come rapid expansion until the local organization had grown cumbersome. They decided they needed to halt this trend and reorganize what they had in Santa Clara into smaller, more manageable elements. In addition, the area (by then dubbed "Silicon Valley") was becoming so overcrowded that there were fewer and fewer suitable places for the work force to live.[15] This was of great concern to the founders, especially Grove, who took the lead in shaping what came to be known as the Intel culture.[16]

These perceptions dictated a move away from the valley. But where? For a number of reasons, eastern Washington County in Oregon appeared the best choice. It was close to the Bay Area, and therefore it was readily accessible to its parent, its suppliers, and its biggest customers.[17] It had a stable and predictable supply of energy and water. And it was near Portland, which would add a touch of urbanity to attract young professionals.[18]

Furthermore, 1974 was perhaps a high point in the state's funding of higher education.[19] Although a downward spiral in the higher-education budget began in the next year, to outsiders such as Intel's leadership, Oregon's colleges and universities looked as good as they ever would.[20] This was important to a company competing in an arena in which the work force constantly had to retrain in order to stay ahead.

As for land prices and construction costs in Aloha, where Intel eventually settled, they were far below those of most places in the nation, including those of the Bay Area.[21] Taken together, these considerations shaped the decision to move to Washington County.

Thus, on 30 January 1974, Eldon Hout, a Washington County commissioner, announced that Intel had taken an option on twenty acres of land at Southwest 198th and Tualatin Valley Highway in Aloha for the purpose of constructing a forty-thousand-square-foot semiconductor manufacturing plant, which in its initial phase of operation would employ four hundred persons.[22] In its fourth and final phase the plant was projected at one hundred sixty thousand square feet, with a labor force of sixteen hundred, making it one of the five largest employers in the county.[23]

The plant was scheduled to open in late 1974, but construction was slowed by a recession that lasted through 1975.[24] By October 1976, however, Intel had hired 125 workers, and in December the plant had begun production on the 16K random-access memory, a product introduced the previous year.[25]

By 1979 Intel had twenty-six hundred employees at the Aloha plant—"Fab 4," as it was known.[26] Despite its slow beginnings, much of the company's expansion occurred in Oregon. In 1978 almost one-fourth of the company's U.S. employees were in Oregon.[27] (Worldwide expansion into Mexico, Malaysia, the Philippines, and Hong Kong and domestic expansion into Arizona and New Mexico was also undertaken in the seventies.)

In September 1979 the company announced that it was going to add an operation for peripherals and instrumentation in Washington County. Pending completion of a new plant in the Hawthorne Farm Industrial Park in Hillsboro, the group would work in leased quarters next to the Aloha plant.[28] The new plant's labor force, said Oregon's governor, Bob Straub, would make Intel the state's sixth largest employer.[29]

In retrospect it is clear that Intel's arrival in Washington County was significant for reasons that went far beyond the immediate economic boost it provided. More important was the invigoration of the county's high-technology industry that the company's presence contributed. By the late seventies, Intel was recognized as a distinguished leader in electronics. Noyce and Moore, in particular, had achieved great things at Fairchild in the fifties and sixties and traced their lineage back to Shockley, in many ways the Columbus of twentieth-century electronics. At Intel, they and others—young, eager, and visionary—had continued to push back the frontiers of the field with breakthroughs such as the 1103 and the microprocessor. And the company's emphasis on research and development has paid off, as an impressive 60 percent of its product line reached market within two years of development.[30]

This strangely contradictory cachet—part elder-statesman solidity and sobriety, part whiz-kid energy and productivity—with a record of real accomplishment made Intel a cynosure in the electronics industry. Some of the focus had by that time shifted to its operations in Washington County, where the company continued to present innovations. Such products as the iAPX-432 family of computer chips, which had the computing power of a large mainframe computer, and the System 86/330, Intel's first integrated microcomputer systems, introduced in 1981, had their roots in Oregon.[31] In Washington County, where Tektronix and ESI had built remarkable but still hardly famous names, Intel drew the attention of the electronics community.

At the same time, the company was largely responsible for the development of what might be called an electronics-industry infrastructure in the county. Whereas Tektronix preferred to build its own components, from cathode-ray tubes to coils and circuit boards, Intel required suppliers.[32] Accordingly, it lured some of its California suppliers—materials firms such as Buehler and the high-tech public-relations firm of Regis McKenna—into Washington County.[33]

Thus, with its large, trained work force, its renown, and its suppliers, Intel's arrival into Washington County expanded the base created in the forties and fifties by the pioneers, Tektronix and ESI; focused attention on electronics; and generally created the conditions under which the industry could, in the early eighties, greatly enhance the enormous growth in the area's high technology.[34]

Yet Intel was not immune to the industry's vagaries. On the contrary, because it was a commodity chip maker, manufacturing in volume for a variety of products (as opposed to custom chip makers, who manufacture special chips for discrete applications), the company was extremely sensitive to the fluctuations of the worldwide electronics market. Consequently, in the recessions of 1974, 1979, and 1983, Intel and its Washington County operations were more vulnerable to the downturns than other firms, although the industry as a whole also suffered. For example, Fab 4, begun in 1974, did not open until 1977, having been mothballed in the interim. In November 1985, an even more severe blow was felt as electronics manufacturers overbought chips by 20 percent and then, assessing glutted inventories, stopped ordering. The situation was compounded by stepped-up competition from Japanese chip makers. In the face of a $4 million companywide loss in the third quarter of 1985, Intel not only announced that it was shutting down Fab 5, where it made dynamic random-access memory chips (DRAMS),[35] but decided to withdraw completely from that market, which had been lost almost entirely to the Japanese.[36] "We simply chose not to be a fringe player in that market any longer," said Grove.[37] As a result, about 450 of the 520 workers at the plant were laid off.[38] Earlier, the company had phased out 75 workers at the Hillsboro plant for the same reason.[39] In 1981 the company had mothballed a plant under construction at its Hillsboro site.

By early 1986, Intel had thirty-two hundred employees in Washington County, the fewest since 1982, when there were about four thousand on the payroll.[40] Yet despite the perilous markets, the company had accomplished much in Washington County, a fact that was earlier symbolized by its first annual meeting held outside California on 16 April 1980. At a Beaverton hotel, Moore, Noyce, and Grove emphasized to the nine hundred persons in attendance Intel's accomplishments in Washington County and their optimism for its future in the state.[41] Even in late 1986 that optimism held, for, in addition to the iAPX-432 chips and the system 86/330 computer, the Washington County plants had developed and produced the 216A, a 64K RAM, a "Super-view" 64-kilobyte office computer called the System 310, a high-performance scientific computer called the iPSC-VX, and an add-on printed circuit board that greatly enhanced the memory of a personal computer. For the first time, the marketing of these products placed the company in the position of being a retailer.[42] The iPSC-VX, marketed in the spring of 1986, seemed to sum up Intel's contribution to Washington County's history of high technology. Its development and recognition by consumers, the *Oregonian* reported, woke the industry to the fact that in light of other, similar machines made there, the county had become a "hotbed of scientific computing research and development."[43]

FLIGHT DYNAMICS

In 1977, one year after the coming of Intel, a small company was founded in Hillsboro to manufacture radio-controlled model airplanes. Five years later, Flight Dynamics was producing a rather different type of product for the aviation industry. The company was established by Robert L. Carter, once a flight-test engineer for the Boeing Company, who had gone on to be one of the original founders of Floating Point Systems.[44]

At Flight Dynamics, Carter's interest soon turned from radio to another type of flight control. He became interested in head-up display (HUD) systems, which were then being used only by the military in pilot training. The HUD is designed to enable airplane pilots to look at a panel display in their line of sight (thus "head up"), rather than down at instruments when landing or taking off. Flight Dynamics intended to use laser-produced holographic optics to give the pilot guidance and control information. The panel display would include air speed, altitude, horizon, a synthetic runway, and other data, all of which would be updated instantly as flight conditions changed. In fog or sudden alteration of wind direction, for example, the panel display would enable the pilot to locate the center of the airport runway. If on takeoff one engine failed, the HUD would furnish the pilot essential information to attempt to land the plane safely.

Flight Dynamics engineers became interested in the HUD in 1979 and after two years completed its prototype holographic-optical system. As one writer explained: "A hologram is a three-dimensional picture projected onto a plate of some kind without using a camera. The Flight Dynamics holographic combiner consists of a wave length or color selective mirror within a thin transparent gelatin film. This is sandwiched and sealed between two pieces of optical quality glass." One of the engineers said that their greatest concern was "that if the machine flakes out that it either shuts off completely or that it notifies the pilot that something is wrong."[45]

Capital for Flight Dynamics' formative years came from a variety of sources. Carter invested $1 million of his own seed money; Federal Express, one of the company's first customers, provided $90,000; and Questech, a New York venture-capital firm, contributed $500,000. In May 1981 the company became a publicly owned corporation when it offered 1.5 million shares over the counter. The $3.5 million result of the offering gave the company financial stability until the first sales occurred.[46]

Besides Federal Express, Flight Dynamics' first customers in 1981 were Boeing and NASA. In 1982 the company announced its first major sale—ten of its HUD systems, to Airborne Express, for $2 million. This success led to a major investment in late 1982 by Pacific Telecom, a subsidiary of Pacific Power and Light, which agreed to buy 40 percent of Flight Dynamics' stock for $5.5 million.[47]

These financial achievements were enhanced by two significant developments in 1985. On 22 April the company received Federal Aviation Administration certification of its HUD system as a landing device for the Boeing 727 in difficult conditions where runway visibility could be as low as 700 feet. The federal stamp of approval meant that the company had the potential to expand its market enormously.[48]

In August, Flight Dynamics acquired a company, in Salem, Oregon, ARNAV Systems. Begun in 1970 as Morrow Electronics, ARNAV was one of the largest producers of long-range navigation (LORAN) equipment. First developed in the forties for marine navigation, LORAN became popular in commercial aviation. When the company was reorganized in 1982, ARNAV developed a receiver that provided many functions previously too expensive to incorporate into aircraft for general (private) aviation. The ARNAV receivers enabled pilots to fly a more direct course to a destination rather than zigzagging from one very-high-frequency radio station to another at a high cost in fuel and time. The ARNAV acquisition was financed by an agreement with Pacific Telecom that resulted in its increasing its share of Flight Dynamics' stock to 82 percent. The wisdom of this acquisition was shown by the contrasting sales figures for 1985 of LORAN ($3,378,000) and the HUD ($505,000).[49]

In 1986 Flight Dynamics obtained a $1.1 million contract from Federal Express to supply the equipment and training needed for that company to include the HUD systems in its Boeing 727 flight operations. Other contracts that year included the company's first international agreement (with the People's Republic of China National Aerotechnology Corporation) and an order with the Lockheed-Georgia Company for its third HUD system to be used in its High Technology Test Bed military transport.[50]

MENTOR GRAPHICS

In late 1985, only four years after its birth, Mentor Graphics in Beaverton, one of a handful of companies in the United States manufacturing computer-aided engineering equipment, held a 35 percent share of its market. This figure represented sales of $87.9 million and preeminence over its two rivals, Daisy Systems and Valid Logic.[51] The demand for Mentor's IDEA series of engineering workstations had made its product line an industry standard and the company virtually an overnight success story. By the end of 1985, Mentor was Oregon's second largest indigenous high-technology employer, after Tektronix. Its rapid rise surprised even its founders. Thomas Bruggere, the company's chairman and chief executive officer, had projected sales of $5 million for 1983. The next year, he planned to double that figure, and in 1985 he hoped sales would reach $24 million.[52] Mentor's position at the end of 1985 was the more unexpected in light of the deep recession besetting the electronics industry.

However, the company's rise was the result of a thoroughly calculated and rationally planned exercise in starting a firm, making a product, and winning customers in the superheated worldwide electronics environment of the eighties. In 1976 Bruggere, a combat veteran of Vietnam and a former software manager at Tektronix, had decided that he wanted to learn the industry and advance through its ranks faster than was then possible at Tektronix.[53] To do so he would found a company that would grow into a major publicly held corporation. To help him, he recruited top personnel from Tektronix, including Gerard Langeler, who was marketing manager in the Information Display Division, and David Moffenbeier, a certified public accountant in the budgeting department who had worked previously at a large accounting firm.

Initially Mentor's founders had no idea what product they wished their company to manufacture and market. They only knew that during their tenure Tektronix was not exciting for young entrepreneurs and that innovation, aggressiveness, and imagination in developing new products were not Tektronix's foremost qualities in the late seventies.[54] In forming a new entity, they wished to elevate those traits to primary status.

Consequently, in looking for marketing opportunities, they searched for needs within the engineering community, examining trends and problems. Eventually it became clear that the greatest prospects lay in the field of computer-aided engineering (CAE). Applied to the electronics industry, a CAE system provides engineers with the capability to "capture, analyze, verify, lay out, test and document complex electronics products," largely by replacing paper and pencil with a computer and sophisticated software.[55] Moreover, the system allows a manager to coordinate and monitor the activities of teams of engineers. In an integrated and automated system, CAE can be the key to advances in productivity while aiding innovation through the use of a single design path from concept to prototype.[56]

In reality, the choice of CAE was not so easy, for another opportunity in a related field, computer-aided manufacturing (CAM) workstations, presented itself. CAM offered a market potentially four or five times larger than CAE. However, the new company's engineering talent was more familiar with CAE, and in the end it decided to stay with the smaller field.

In a step that would prove critical from the standpoint of marketing, Bruggere and his partners continued to survey engineers about the most desirable features of a CAE capability. At the same time, they began to approach venture capitalists for backing. The financial community was apparently pleased with Mentor's blueprints, because in the initial round of approaches funds amounting to more than $1 million came from Venrock Associates, Sutter Hill, and Greylock, firms that had previously subsidized companies such as Intel and Apple Computer.[57] In July 1982, $2 million came from Hambrecht and Quist; L.F. Rothschild, Unterberg, and Tombin; and Lamoreaux and Glynn. In April 1983 the original six investors and nine new ones added another $7 million to Mentor's treasury.[58]

What had apparently impressed the financiers was the combination of the managerial and engineering team, the product, and the market potential that Bruggere and his cofounders had assembled. This was reflected in what by all accounts was a brilliant business plan. By 1983, when the bulk of the company's venture capital was secured, Mentor's leadership was widened to include a variety of skilled scientists, engineers, and managers, including Ning Nan. A former high-technology entrepreneur, Nan had also been a member of IBM's Corporate Technical Committee. As the founder of California Automated Design, Inc. (CADI), Nan was made vice-chairman and manager of the Automated Layouts Division when his company was acquired by Mentor in 1983.[59] Stephen Swerling, an electrical engineer and a cofounder of Mentor, had been involved with new-product development at Tektronix and became vice-president for Mentor's Computer Systems Division. Michael Feuer, chief scientist at Mentor, also came from CADI. With others whose experience in electronics at companies such as Hewlett-Packard, AT&T, IBM, and Sparcatron added years of experience and a diversity of knowledge to the Mentor profile, they reassured potential backers.[60]

The founders' choice of product and manufacturing strategy were similarly appealing to the financiers. The CAE workstation constituted a major advance in rationalizing and optimizing efficiency in electronics design. The Mentor package of software and hardware promised to streamline all aspects of product development, automating each isolated process and then integrating it with the others.[61] The CAE workstation promised great upswings in productivity and originality of design. Such potential would be hard for the engineering community to resist.

In manufacturing strategy, the company's key decision was not to build its own hardware, but to combine its specially created software with the Apollo Computer Domain workstations and the Aegis operating system. Although this was risky because Apollo was also a start-up, the decision relieved Mentor of huge expenditures for research and development. Moreover, the company was avoiding the problems associated with obsolescence and compatibility of hardware of various generations.[62] In the future, customers would not be wed to one line of additions only, an important point in initial sales, when they would be contemplating purchases ranging in price from $20,000 to $100,000.[63]

In terms of market potential Mentor had entered an almost unexploited field, where there would be a huge demand from the world's system designers, who numbered from three hundred thousand to four hundred thousand, and designers of integrated circuits, the early users of CAE devices, who numbered between five and ten thousand.[64] In 1980 this great market in the semiconductor, aerospace, computer, consumer-electronics, and telecommunications industries generated a $150 million demand and by the end of the decade would constitute a market valued at $1.5 billion.[65] Later investors were impressed that Mentor was a company designed to be driven by its markets.[66] Rather than being an engineer-led company that existed to be first in the marketplace with the newest and most advanced products, it intended to let its markets tell the company what they wanted and then design products to those specifications.[67] The strategy was acknowledged to have worked with great success in recent years for IBM.

Bruggere, Langeler, and Moffenbeier had formed their company in April 1981; by 15 May, they had settled on their product and begun to acquire financial backing.[68] By November 1982 Mentor had shipped its first products. Yet the first CAE workstations had already been in the hands of some customers for several months. Daisy Systems, a California firm founded in 1980, had begun manufacturing and shipping them in November 1981, followed quickly by Valid Logic Systems of San Jose. In 1982 both Daisy and Valid Logic led Mentor in that year's share of the $12 million market.[69] In 1983, however, Mentor passed Valid Logic, and in 1984 it outshipped both rivals in what had grown to be a $250 million market.[70] The company's early decisions with respect to marketing and manufacturing had thus almost immediately bore fruit.

Whereas Mentor had chosen Apollo hardware, Daisy and Valid Logic had decided to build their own. Mentor's choice was what one management analyst called a "bet-your-company decision."[71] In 1983, Apollo was itself essentially a start-up company, but it manufactured one of the most advanced computers of its kind. Several Mentor customers acknowledged that it was the Apollo hardware that convinced them to buy Mentor over its competitors.[72] Apollo offered other advantages to Mentor, not the least of which was its network of funding sources and its trailblazing in the market to which Mentor wished to sell. As one observer put it, in a kind of symbiotic relationship the two fledgling companies clung together like shipwrecked sailors in choppy seas.[73] It was unquestionably a source of strength as Mentor faced its early challenges.

It seems clear that Mentor's intensive questioning of engineers about what they wished to see in CAE software paid dividends. Both of its major competitors had in essence "productized" existing software packages. Valid Logic had simply used SCALD, which was developed at Stanford University, while Daisy had "front-ended," or added steps to, a computer-aided design package.[74] Because CAD is a different process from CAE, the market saw Daisy's software as a hybrid, Valid Logic's as serviceable but not entirely responsive to user needs, but Mentor's as addressing its market's special requirements.

Contributing to Mentor's rise was the responsiveness of its sales and service force to customer needs. When the software didn't work for some reason, a Mentor representative was usually available to solve the problem in a matter of hours. To businesses in which thousands, perhaps millions, of dollars are regularly at stake, this speed was impressive.[75]

To reach its small but definitive lead in the CAE market, Bruggere had made fortunate choices. There were, however, other difficult options ahead, the most critical of which concerned the use of Apollo hardware. The question was complex because both IBM and Digital Equipment Corporation had entered the race for a share of the CAE hardware market. Mentor's management recognized that to retain its share of the market, the company might be forced to link its software to all three—Apollo, IBM, and DEC.[76] Eventually, the company chose to stay with Apollo and to add IBM. Daisy and Valid Logic, meanwhile, had agreed to use the DEC platform, the Micro VAXII, a workstation with great early support in the electronics industry. Mentor's decision was justified when Apollo outdid DEC and IBM in the workstation market. Daisy, however, suffered from the long delay attending the switchover to the DEC-based workstations, which crippled its sales. The company never recovered.

Also critical to Mentor was its decision to acquire CADI in September 1983. In 1982 the latter was a start-up company in Santa Clara, developing software programs to simplify the design of gate-array or standard-cell circuits. Although Mentor wanted to service that segment of the CAE market, CADI had stolen a march on it. In a defensive move, Mentor's management quickly decided to acquire CADI, rather than try to compete with it. The CADI acquisition was only a qualified success.[77] Although the new enterprise immediately added the capability Mentor wanted, as well as entrance to the market CADI had already pioneered, market growth was not nearly so rapid as Mentor had forecast.[78] It is speculated that the new company was overpriced;[79] in addition, the attempt to form two corporate cultures into one cohesive working unit proved difficult.[80] (By mid-1986 Nan had resigned from the company, after a year-long sabbatical.)

As Mentor has matured, so has its market. With the market's expansion has come competition from many quarters. Well-established firms (such as Intergraph of Huntsville, Alabama; Computer-vision of Bedford, Massachusetts; and Metheus Corporation in a joint venture) and older firms (such as Tektronix and Hewlett-Packard) entered the CAE field.[81] They have provided an important challenge to Mentor. Nevertheless, public faith in the company was affirmed in 1984 when its first public stock offering was judged by the financial community to be the best of the year.

To meet the challenges of its competition, the rapid pace of technological change, and its own internal goals and standards, Mentor's top management has set a demanding internal tone and pace. The firm likes to be "hardworking," to use the CEO's phrase, with lights burning in offices sometimes as late as 2:00 A.M.[82] The bottom 10 percent of producers in the company are constantly being identified and weeded out.[83] On the other hand, as growth has come, decisions have been pushed outward from top management so that those closest to the market have ultimate influence on the company's direction in its product lines. Divisionalization was instituted early to create the greatest independence and creative freedom for the technical staff.[84] This, Bruggere suggests, has fostered an atmosphere like that at Hewlett-Packard, a highly successful company that he openly admires.[85]

As in most aspects of the electronics industry, Mentor's rise is an unfinished story. Capitalizing on opportunity and circumstance, brilliance and fortune, the company and its founders have made names in a field they themselves partially created. To continue to flourish, those qualities and that luck must be sustained.

METHEUS CORPORATION

Metheus Corporation of Hillsboro is another computer-graphics corporation. It was founded in 1981 by Gene Chao, an electrical-engineering graduate of Berkeley, an Apollo space-program employee (he constructed circuitry for a radar antenna that was left on the moon), and former head of Tektronix's research group. Metheus is a designer and manufacturer of graphics systems for the computer-graphics market. In contrast to its Oregon rival, Mentor Graphics, its history is not an unadulterated success story, but a series of advances and retreats.[86]

From the beginning, Metheus produced computer-graphics systems for original equipment manufacturers, which add the systems to their own before selling them to the final, or end, users. A year after its founding, Metheus shipped its first product, the Omega 400 Display Controller.[87]

On 4 June 1984 the CAE industry was hit by a bombshell: tiny Metheus formed a joint venture with massive Computer-vision of Bedford, Massachusetts. The intention was to create a new company, combining Metheus's CAE capabilities with Computer-vision's strengths in CAD and computer-aided manufacturing, or CAM. Their product was to be a single system combining all three computer-aided processes.[88]

The president of the new company, Metheus-CV, was James Towne, the former president of Metheus. His firm was created through some complex financial arrangements. Computer-vision lent $10 million, with the right to waive repayment in return for obtaining a majority of Metheus's stock. In addition to cash, Metheus obtained national exposure and a broad marketing network from the deal, while Computer-vision got electrical engineering software and a full-color workstation.[89]

News of the joint venture was received with mixed responses from competitors and financial analysts, but in February 1986 those on the negative side were proved right when the firm collapsed. As a result Towne resigned as president, the sales and administrative staffs were laid off, and the company became a research-and-development branch of Computer-vision. The failure was due to friction between the new firm and Computer-vision, the limited and tightly held CAE market, and the possibility that Computer-vision wished to develop CAE internally. Clearly Metheus-CV's financial goals were not reached: instead of earning $30 million in 1984 and $66 million in 1985, the peak year was $12 million and sales stood at a mere $10 million at the time of collapse. Meanwhile Metheus pressed forward alone, attempting to ride out the blow.[90]

Despite these vicissitudes, Metheus's revenues had grown to $19 million by fiscal year 1985 from the production of graphics-display controllers that allowed fast drawing and redrawing of images on the screen, high resolution, and bright colors. The company had also made several improvements in the Omega 400, and in April 1985 it had moved on to a new generation of controllers, which improved performance by adding a graphics preprocessor to the existing display controller.[91]

PLANAR SYSTEMS

Another notable new firm, Planar Systems, represented a technological innovation and was the first spin-off that Tektronix has supported with its own venture capital. The company was founded by James W. Hurd, the former head of the solid-state laboratory at Tektronix, to manufacture electroluminescent (EL) display panels.[92]

Display panels are the "screens" of television sets, computer terminals, watches, and other products. The first screens were illuminated cathode-ray tubes (CRT), but by the seventies engineers were working to develop more efficient replacements. The contenders included flat-panel displays based on gas plasma, liquid crystal, and electroluminescent materials. The advantages of flat-panel displays over CRTs are that they are lighter, more compact, and more durable and use less power. CRTs, however, were initially easier and less expensive to manufacture. Among the three flat-panel methods, plasma displays for screens measuring more than six inches diagonally were less likely to flicker than the electroluminescent displays, while displays based on liquid crystal required less power than gas plasma. EL displays, because of their solid-state construction, were easier to read than liquid-crystal displays, although they required more electric power. Because of their brightness, contrast, and viewing angle, EL displays were considerably superior to the other display technologies. Because of their cost, however, before 1983 flat-panel displays were used only for readout devices (as in watches) and specialized, limited markets such as scientific instrumentation.[93]

In May 1983 Planar Systems was formed in an attempt to broaden the market for electroluminescent panels. By that time, knowledge of electroluminescence (the production of light by a phosphor under the influence of an electric field) had been around for some time. In 1934 the luminescence of zinc sulfide in an electrical field was discovered, and it was followed by several years of low-key research. In 1950 the GTE Company attempted to place four-by-eight-foot EL lighting panels in house walls, but with little success. By the seventies and early eighties various companies, including Sharp, IBM, GTE, Aerojet-General, Matsushita, Rockwell, and Tektronix, were working on electroluminescence. In 1974 Sharp introduced a design that first showed that EL displays were commercially viable. At Tektronix, Hurd and Chin King came up with new technology for the displays, but Tek had no practical use for it and was preparing to cancel the program.[94]

Hurd and King, convinced that their discoveries could replace CRT displays, then approached Tektronix for permission to establish a friendly spin-off company to which Tek would obtain entry. They pursued their request for over a year in the face of opposition from middle management, until Vollum and Chief Operations Officer William Walker at last approved it. The initial funding for Planar came from Tektronix, with additional capital derived when the new company's first product was sold. The deal that Planar negotiated with Tektronix included exclusive rights to all technology developed by Tektronix in the field of EL displays; several million dollars in lease guarantees; a short-term loan to Planar until equity financing could be arranged; and access to Tek's EL facilities until its own could be built.[95]

The new company was born into a competitive world dependent

on a military market that required compact, durable, and portable gear. By 1985 the major manufacturers of EL displays were Sharp (Japan), Planar, Lohja (Finland), Sigmatron Nove, and Phosphon (Great Britain), while some of the competitors for the military market were Sharp, Photonics, Luckey, Ohio, and Electro-Plasma. Planar also started in a lower market category than those firms. In 1981 the plasma-display market totaled $16 million, while the EL display market was nonexistent. On the more positive side, Planar began corporate life with a far cheaper product than plasma displays.[96]

Planar's first major commercial customer was Grid Systems, a California manufacturer that used its display on eleven-pound computers. The U.S. Army also used a Planar display for a portable field terminal. But success could not be taken for granted. For example, all EL companies were competing to produce color panels that would replace monochromatic screens. (In this regard, Planar was helped by a $1 million contract from the army.) Other challenges were to reduce power consumption and make panels more readable in sunlight. Furthermore, CRT manufacturers were trying to make their terminals more compact and less power consuming. Sony produced the first flat-screen CRT television set in 1985, but other multinational firms, such as RCA, Matsushita, and Siemens, were also committing their resources to similar improvements.[97]

Planar's competitiveness in these international markets was shown in late 1985 and early 1986. In December 1985 RCA announced a $10 million investment in Planar, with an agreement to exchange technology and operations regarding information and graphic displays. As part of the agreement, two RCA officials joined the Planar board of directors.[98] This apparent coup, however, rapidly turned into a mixed blessing. Within days of the agreement, RCA was acquired by General Electric. General Electric's plans for the display components it developed were the opposite of RCA's. By mid-1986, the relationship between Planar and RCA had deteriorated badly. Intense negotiations between General Electric and Planar were required to untangle the original agreement. By the end of 1986 it had been decided that General Electric would sell back to Planar the stock that RCA had originally purchased in the company.[99] The whole episode resulted in developmental delays for Planar.[100]

In the midst of these difficulties, a more positive event occurred. In May 1986 Data General Corporation gave Planar its largest sale when it agreed to incorporate the Oregon company's EL display in its new Data General One Model 2 lap-top computer. This gave Planar an entry into a corporation that had 18 percent of the lap-top market and the first EL display on a portable IBM-compatible computer. Although the future was still uncertain for Planar in mid-1986, the company had made much progress in a short time.

LATTICE INTERNATIONAL

In the silicon-chip area, Washington County's major new manufacturer by 1986 was Lattice Semiconductor Corporation, founded in 1983 by Rahul Sud, a former circuit designer at Intel. The company began with high hopes of manufacturing chips by using a new technology, but management shake-ups and delays in introducing the product and meeting sales targets removed some of the company's initial luster. The older technology for manufacturing silicon chips was a bipolar one, while Lattice's process used the newer complementary-metal-oxide-semiconductor (CMOS) technology, which required less power and less heat.[101]

The company's initial goal was to reach four markets with four types of CMOS chips. One was a random-access memory chip, which is expensive but does huge calculations quickly. Another was reprogrammable-logic devices, which can be programmed and reprogrammed electronically to do a variety of tasks. The third was a digital signal process which helps to convert sound to a digital format, while the fourth was an E^4 PROM, an electrically erasable programmable read-only memory. Capital for Lattice came mainly from Louisiana-Pacific Corporation and its president, Harry Merlo; a consortium of Norwegian businesses; and Floating Point Systems, as well as other local businesses and individuals. Louisiana-Pacific and the foreign investors were interested in providing a conventional source of funding, but Floating Point's concern was somewhat different. That company became interested in Lattice when it decided to use CMOS chips in its computers. C. Norman Winningstad, impressed with Lattice and its proximity to Floating Point, took the idea of investing in the start-up company to his board of directors, whose members initially resisted it. In time, however, he convinced them by making a large personal investment himself. Other potential investors remained reluctant because of Sud's lack of management experience, the variety of markets he was trying to reach, and Lattice's intention to build its own chip-manufacturing plant, a costly venture for an untried company.[102]

Eventually, the company decided to subcontract chip-manufacturing space in California and Japan. It produced its first chips in 1985 and gained increased prestige in 1986 when it was chosen as a supplier for the military. A shortage of cash required further investment. When that was not forthcoming, Lattice had to reorganize. The company dropped three of its four products, concentrating on the generic-array-logic (GAL) product, which had a short time from conception to marketing and relatively little competition. Based on a successful focus on the GAL product, Lattice survived the reorganization and emerged to healthy growth.[103]

JAPANESE FIRMS

A relatively new aspect of the Washington County high-technology community is the presence of Japanese electronics companies. Their arrival occurred within a few months' of each other because of a combination of local and international developments, some of a long-standing nature, and some that were entirely novel. Beyond Oregon's borders, the most important developments were political and technological.

In January 1984 a federal district court judge ordered the breakup of the American Telephone and Telegraph Company into seven regional Bell operating companies. This decision opened the American telecommunications business to competition and resulted in the development of many new markets, including private branch exchanges and key systems for the Bell operating companies; fiber-optic and microwave equipment for telephone companies; digital facsimile machines for facsimile-transmitted mail service companies; and cellular radio equipment. The great technological development was in fiber optics, the use of glass wires and tubes to transmit the human voice by laser light beams, a far speedier means of communication than the old technology of copper wires operating on electrical impulses.[104]

Given Japan's economic and technological prowess, it was natural that it would seize opportunities presented by divestiture and fiber-optics development to penetrate U.S. markets. One way to accomplish this was to open manufacturing plants, marketing distributorships, and research laboratories in the United States, rather than to export goods to it. An advantage to this approach was the avoidance of restrictions on imports, which had been imposed in the late seventies and early eighties as the American economy reeled from an enormous increase in Japanese goods. Another advantage was that of geographical proximity: manufacturing and distribution facilities would be closer to U.S. markets, reducing transportation costs and furnishing better customer service. Moreover, in the early eighties the dollar was strong compared with the yen, a condition providing a high rate of return for investment in the United States. This trend has been confirmed by the fact that Japanese companies invested between $2 million and $3 million abroad annually in the early seventies but raised the amount to between $8 million and $9 million a decade later.[105]

NIPPON ELECTRIC COMPANY

In April 1984 Oregonians began to benefit from these economic conditions when the Nippon Electric Company (NEC) purchased a 210-acre site in Hillsboro for a plant to manufacture fiber-optic communications equipment. NEC had been founded in Tokyo in 1899 as Japan's first joint-venture company (with Western Electric). Later it expanded, until by 1984 it was one of the world's largest multinational electronics companies. Company officials were interested in a greater market share in the growing telecommunications business. Accordingly, NEC's communications business group designated a subsidiary, NEC America, to build and operate a plant in Oregon. At the time, NEC America had been in business for twenty-one years and was operating plants in California, Virginia, and Texas.[106]

On 1 October 1985 NEC's Hillsboro facility began operations. The plant, which cost $30 million, employed one hundred men and women to manufacture fiber-optic transmission systems, digital microwave radio equipment, data modems, and cellular mobile telephones. The Hillsboro plant (the ninth built by NEC in the United States) was one of the most highly automated electronic assembly plants in the United States. What struck public fancy about the plant was its two automatic delivery machines, soon nicknamed Bonnie and Clyde, which delivered parts to the workers on computer commands. Other machines made inventory searches, assembled components, and tested products.[107]

EPSON CORPORATION

The second Japanese company to come to Oregon, Seiko Epson Corporation, was established on 1 November 1985 through the merger of Suwa Seikosha Company and Epson. The latter was a major manufacturer of computers and peripherals—printers, disk drives, and liquid-crystal-display products. The firm's American marketing and sales subsidiary, Epson America, had earlier set up an advanced product-development group in California as well as a marketing network. It also had a software development group.

By 1984 Epson was ready to open manufacturing facilities in the United States to complete the third leg of its "triangle theory model," which involved establishing its plants where they could best serve the needs of their American customers. After deciding to manufacture in Oregon, Epson established a Northwest subsidiary, Epson Portland, to carry on this operation.[108] In October 1984 it announced that it would build a dot-matrix printer plant in the Sunset Corridor.

Epson's roots in Washington County went back to a visit to Oregon in February 1983 by Masakatsu Tomita, project manager for Epson America, who was impressed by the firm's future location in Hillsboro: "Our future location was then a green wheat field and a big Douglas fir forest," he said. "It was a sleeping beauty—peaceful and beautiful enough to fascinate my imagination." The next year Epson and Standard Insurance Company worked out an agreement whereby Standard developed the Tanasbourne Commerce Center for the Japanese firm.[109]

When the plant was announced, corporate officials said that the printer plant would be followed by a factory to build personal computers and perhaps another to build semiconductors. Initially, production consisted of assembling components received from Japan into the finished product. A Japanese business newspaper reported that the company hoped to "make the Portland plant a production base for all kinds of information equipment for sale in the world's markets" and indicated the strong possibility that production would "greatly increase in the future." The plant itself was partially financed by an $8 million state industrial reserve bond.[110]

FUJITSU

Fujitsu was the third largest Japanese company to settle in Washington County, when it made known in September 1984 that it would build a plant at an unspecified site there. In January 1985 it announced the acquisition of a 120-acre site in Hillsboro, which was later increased to 145 acres.

The parent company traces its roots to 1896 but took its modern name in 1935. By 1984 Fujitsu America (its subsidiary, founded in 1968) was planning to increase its American sales by 50 percent. Its aim was not only the U.S. computer market, but also the recently deregulated telecommunications market. Its Hillsboro plant, a $40 million building that occupied 144,000 square feet, opened on 7 April 1986, employing forty-six full-time workers. They produced 10 1/2-inch disk drives, the information-storage components of computers.[111]

OTHER COMPANIES

NEC, Epson, and Fujitsu announced openings in Oregon within a few months of each other. However, other opportunities for Japanese business were available in Washington County. For example, the Shimizu Construction Company designed and constructed the Epson plant in Hillsboro, and Kyocera Corporation, a Vancouver, Washington manufacturer of chip capacitors, invested in Synatron Corporation of Tigard, a manufacturer of small, high-speed electric motors.[112]

JAPAN'S MOVE TO OREGON

Just as Japanese firms came to the United States for several reasons, why they chose one part of Oregon is also a complex matter. One theory in 1984 was that Oregon was the first state to repeal its unitary tax (in a special July 1984 legislative session). There is certainly some force to this suggestion, but Epson and NEC had made their commitments to the state earlier. Epson had been talking to state officials for two years concerning repeal of the tax, and the governor, Victor Atiyeh, had led several trade missions to Japan. It was during one of these visits that Fujitsu announced that it would build a plant in Oregon.[113]

The governor's travels to Asia had extended over several years, earning him the nickname "Trader Vic." In 1979 he had journeyed to Japan, Korea, and Taiwan; in 1983 he had gone to Japan again and had revisited that country the same year and in 1984, 1985 (twice), and 1986. Japanese firms had also scouted Oregon. In 1984 and later in 1986 representatives of the Japanese Federation of Economic Organizations (the Keidanren) visited the state to get a firsthand view of its possibilities for investment.[114]

The coming of NEC to Oregon furnishes an instructive case study of Oregon's pursuit of Japanese industry. In October 1983 the Japanese firm confidentially contacted the state Economic Development Department to request site and economic information. The department, rather melodramatically, gave this assignment the code name "Cancun." In the next month, NEC America teams made two visits to Oregon to review sites suggested by the department. In February 1984 NEC officials came to the state to discuss specific properties and picked a real estate agent, Coldwell Banker, to help choose a specific site. Later in the month, Atiyeh invited NEC America's president to meet him. Other conferences between state and corporate officials followed in March. By April, NEC America had chosen its law and engineering firms. Early that month, six months after its first show of interest, the company informed the state of its decision to settle in Oregon.[115]

To some extent, history also played a role in bringing Japanese business to Oregon. The Japanese presence, at least at a modest level, was an old one. Moreover, Japanese firms were aware that the giant American electronics firms Hewlett-Packard and Intel had both opened Oregon plants in 1976. Government officials in the state were helpful. NEC opted for Oregon rather than California, not only because California had a unitary tax while Oregon did not, but also because of the attitudes of state, county, and city officials.

One manifestation of Japanese interest in Oregon was the opening of a Japan Representative Office in Tokyo in September 1984. In it, permanent representatives of the Oregon Economic Development Department provided services for both Oregonian and Japanese businesspersons, from identifying potential suppliers and customers to providing cultural information about the two jurisdictions and scheduling appointments with government business officials.

In less publicized ways, the state also contributed to the process. For example, the Economic Development Department helped locate a new school building in Washington County for the Saturday School attended by Japanese children to maintain their language skills and learn the advanced mathematics they would need when they returned to Japan.[116]

Beyond the purview of Oregon's government officials lay other attractions for Japanese investors and manufacturers: a hardworking labor force, relatively cheap industrial land, inexpensive residential housing, relatively inexpensive hydropower, a good elementary and secondary school system, and a "critical mass" of engineers. In addition, the Portland area was only ninety minutes by air from California's Silicon Valley, where Japanese firms had headquarters or plants, and there were nonstop commercial flights between Tokyo and Portland. And, there was what one Fujitsu officer described as "a pleasant lifestyle." Western Oregon's countryside resembled much of Japan. A popular Japanese television series, "From Oregon With Love," had dealt with the adventures of a young boy in Oregon. For its part, NEC had learned something of Oregon in 1982 and 1983, when it had sponsored a Davis Cup tennis competition in Portland. For whatever reasons the Japanese made their decisions, their presence has enriched not only the state's high-technology community, but the lives of all Oregonians.[117]

6

The Supercomputers

IN THE MID-EIGHTIES supercomputers became both a public byword and a vital ingredient in the Washington County high-technology world. At its simplest, a supercomputer is a computer with a large capacity and tremendous speed. It is generally identified with Cray Research of Minneapolis, which placed its Cray I machine on the market in 1976. The Cray I could perform 160 million arithmetical operations per second. By comparison, the typical IBM large mainframe computer could perform 7 million. (However, Crays were expensive. In 1986, they still cost at least $5 million each. Not surprisingly, only 110 had been sold by that date.)

The importance of supercomputers is that they make possible a range of mathematical models that are far beyond the reach of conventional computers. The range is so vast that according to a local journalist, Nobel laureate Kenneth Wilson "compares the development of supercomputing to the Copernican revolution." Obviously, such a technological revolution also suggests a commercial revolution, and in 1986 one financial analyst estimated the world market for supercomputers at $7 billion and growing at the rate of 20 percent a year.[1]

Both old and new Washington County firms have participated in the supercomputer revolution. For most of them, this participation came with the addition to supercomputing of parallel processing, which enables computers to work faster by dividing their problems into components, all of which can be solved at the same time.

FLOATING POINT SYSTEMS

Floating Point Systems had the most spectacular machines, a result of its desire to build the world's fastest computer. The computer originated at a conference at Salishan, Oregon in 1982 when Kenneth G. Wilson, a Nobel Prize laureate at Cornell University, declared he would buy such a machine if one were manufactured. Floating Point took up the challenge which resulted in the FPS-T Series. The design and prototype of the product were both made by Computer Systems Architects, a firm in Provo, Utah.

Meanwhile the national government gave dramatic recognition to supercomputing when in April 1985 the National Science Foundation announced that Cornell had been designated as one of four national supercomputer centers, which it called the Center for Theory and Simulation in Science and Engineering. In May the center, headed by Wilson, began operations with an IBM 3081 hosting three FPS-164 computers.[2]

Introduced to the public on 2 April 1986, Floating Point's T Series was designed to compete in speed with Cray operations but at a much lower cost. Its smallest configuration could do 128 million arithmetical operations per second, the largest, 262 billion. Each subsequent configuration doubled the performance of its predecessor. For the layperson all this speed was mind-boggling. (One writer said that a T-262 could do in one second what would require an individual 35,000 years with pen and pencil.) Still, the machine was not necessarily profitable. Not only was there competition, but the market was limited. Only major business, educational, and military institutions would require supercomputers.[3]

In December 1985, before the T Series had been officially announced, Congress passed a statute giving the Defense Department $10 million to purchase a supercomputer to the standards of the Floating Point machine. Competition for this lucrative market promised to be fierce, although Floating Point would have an edge. Cray promised a machine that could do one billion arithmetical operations per second. Convex, Amdahl, and ETA Systems were all in the race, while rumors of IBM's interest in joining the competition abounded. The Japanese giants—Fujitsu, Hitachi, and NEC—were all there, too. But Floating Point made an excellent start.

The first customer for the new machine was Michigan Technological University's Center for Experimental Computation. The center intended to use the T-20 in the study of rocket propellants and energetic materials in the hopes of making the use of fuels and explosives safer and more reliable. A month after the T Series came on the market, Floating Point shipped its first machine to a commercial customer—Northrop Research and Technology Center in Palos Verdes Peninsula, California—which planned to use it to explore unique parallel-processor configurations for numerical fluid mechanics.[4]

THE COMPETITORS: INTEL, NCUBE, AND SCIENTIFIC COMPUTER SYSTEMS

Some of Floating Point's competitors were also in Washington County. Intel introduced its first supercomputer, the iPSC, in February 1985. Capable of ten megaflops (a megaflop is one million floating-point operations per second), it sold for $850,000. Although Intel's headquarters were in California, the supercomputer was produced in the company's Beaverton Scientific Computers Division. The iPSC consists of one, two, or four computational units, each containing thirty-two high-performance microprocessors. The microcomputers are interconnected in a hypercube topology.[5]

NCube was the second local contestant in the parallel-processing supercomputer contest. Like so many Washington County firms, it was a spin-off from another company, this one founded in 1983 by three former Intel employees, John Palmer, Stephen Colley, and William Richardson. Colley, a graduate of Caltech, had participated in a

NASA program at the Ames Research Laboratory in California while in high school and, as an undergraduate, had worked in Caltech's Jet Propulsion Laboratory. At Intel, he had worked on the first 32-bit microprocessor project. When he left that company to found his own, he took Palmer, a floating-point chip specialist, and Richardson, a designer of silicon chips.

NCube's engineers successfully built a design that integrated on one chip all the logic needed for a fully functioning, 32-bit node inside a hypercube-based, data-flow parallel processor. The machine could do up to five hundred megaflops. The company had tested its machines at Shell Development Company, Oak Ridge National Laboratory, and the University of Michigan. Shell was its first customer and later bought three additional NCube computers, which it donated to Stanford University, the University of California at Berkeley, and Yale University. In addition to the computer, NCube also introduced a parallel-processing board that would be used to turn an IBM personal computer into a multiple-processor parallel system. In 1986 engineers from the Oak Ridge National Laboratory unveiled Hermes II, a robot powered by a 32-bit NCube processor and capable of performing twenty types of physical movement.

As in the case of many firms attempting to bring a new technology to commercial markets, NCube had its competitors. In addition to NCube's, two hypercubes were introduced in 1983—one from Intel's Scientific Computer Division, the other from Amtek of California. In the larger arena of technological competition, NCube was competing not only with hypercube firms, but with bus-based processors made by Sequent Computer Systems and Encore Computer Corporation of Massachusetts and with the whole spectrum of commercial developments in supercomputing.[6]

Another Washington County firm, Scientific Computer Systems of Wilsonville, was not a competitor of Cray but an ally. Founded in October 1983 by Robert Schumann, a former vice-president for sales and marketing at Floating Point, SCS licensed the operating system of the Cray XMP-2 supercomputer to produce a scaled-down version (a minisupercomputer) priced at about five hundred thousand dollars. SCS became the only firm in Oregon, and one of the few in the world, to produce minisupercomputers. The machine would operate at one-quarter the speed of the XMP-2 and cost about one-tenth as much. Such a computer, Schumann anticipated, would find a much larger market than did the Cray machines. SCS hoped to fit in between Cray and slower, lower priced minisupercomputers in the marketplace. It wanted to sell its product to the approximately fifty thousand Cray users who rented or leased time from the Cray systems they could not afford to own.

Schumann's vision was translated into reality by Hanam Potash, a graduate of the UCLA computer science department, who was working for the Burroughs Corporation. He designed the new machine in three weeks' time. After some initial difficulties, Schumann obtained two rounds of venture-capital financing—$2 million in May 1984 from five New York and California firms and $15 million from the initial companies and nine others, led by TA Associates of Boston, one of the nation's largest venture-capital firms.

The first machine, the SCS-40, was announced in March 1986 and donated to the San Diego Supercomputer Center, a research laboratory operated by a consortium of nineteen universities and research institutes. The entry-level model was priced at $595,000. The SCS-40, unlike the Floating Point machines, uses vector, not parallel, architecture. Vector architecture uses single machine instructions that operate on multiple data elements (or vectors) and deals concurrently with various streams of data. By contrast, parallel processing uses multiple processors to execute multiple instruction streams concurrently. By dividing a problem among a number of processors, greater speed can be achieved than if a single processor is used. The weakness of parallel processing is that not every problem can be divided easily into the required segments. Of these two architectural approaches, SCS chose vector for the SCS-40 because of the availability of the Cray operating systems. Regardless of architecture, in late 1985 there were ten American firms in the minisupercomputer market. Whether SCS would survive the competition was open to question; but its fate would not be decided in Oregon, for in August 1986, the company consolidated its Oregon operations at its headquarters in California.[7]

BETTER CHIPS: FROM TRIQUINT, BIT, AND SEQUENT

Speed is the essence of supercomputing, and the essence of all computing is the chip. Chips have traditionally been manufactured from silicon, an element that is a reliable semiconductor. Yet silicon is not the perfect chip material, and the search for materials that are faster conductors has occupied segments of the high-technology industry for years. In the mid-seventies, a new compound, gallium arsenide (GaAs), appeared in the microwave-component market. As a chip material, it offered several advantages over silicon: speed, lower power consumption, higher resistance to electric currents, radiation resistance, and a wider range of operating temperatures. But GaAs wafers, as they are known, are more difficult to produce and thus ten or fifteen times more expensive than silicon ones.[8]

By the eighties expense appeared less of a problem, as potential markets opened in military signal processing, home-satellite television, fiber optics in telecommunications, and supercomputing. These markets inspired several firms—ITT, Texas Instruments, Cray, Westinghouse, Harris, Hughes, and Rockwell, among others—to begin research in GaAs. In addition, the Department of Defense funded pilot production lines with private corporations. Abroad Fujitsu, NEC, Hitachi, and Mitsubishi began developing GaAs integrated circuits. Britain's Plessey Research created a subsidiary to manufacture GaAs chips, as did Ford Microelectronics and Gigabit Logic in the United States. By 1986 several new and old companies had taken the plunge into GaAs.[9]

Among them was TriQuint Semiconductor, an offshoot of Tektronix, founded in 1983. Since 1978 Tektronix had been experimenting with GaAs for use in its own products. These experiments had been successful by 1984, and the company decided to test the market for GaAs chips outside the firm and in response to the many commercial and military inquiries regarding applications. A favorable reception led to the start of TriQuint, through Tek's venture-capital subsidiary, Tektronix Development Corporation. Although founded by Tek, TriQuint anticipated that the parent firm would never account for more than 10 percent of its overall business. TriQuint also sought additional capital beyond Tektronix Development, although the parent company would maintain a controlling interest.

Not all chip-manufacturing firms relying on the supercomputer market used GaAs, however. One that was not was Bipolar Integrated Technology (BIT), the first Oregon-based semiconductor manufacturer, founded in 1983 by George Wilson and other engineers who came mostly from Tektronix. Wilson, who had a master's degree from Seattle University, had worked at Motorola for two years and at Tektronix for seventeen. Most of his career had been concerned with bipolar integrated circuit design and management, a project he had started at Tek.[10]

At BIT he planned to manufacture a new type of bipolar silicon chip. By 1983 chip manufacturers were producing both emitter-coupled logic (ECL), bipolar-process chips, and metal-oxide semiconductor chips (MOS). Each had advantages over the other. The ECL chips were faster and more powerful than MOS, but the latter used less power. Wilson believed that he and his engineers could design chips that would reduce power consumption, increase density, and achieve full ECL speeds. These were large claims. Even by 1985 he told an interviewer: "If I were you, I wouldn't believe them either."[11]

If BIT could produce, however, there were large markets (one estimate was of $1 billion annually by 1990) awaiting—not only for supercomputers, but for military hardware, radar equipment, and medical imaging machines (CAT scanners). The chips' job would be to transfer data into video-screen shapes and forms. To reach this lucrative market, BIT needed capital. The company's chief financial officer, Ken Giles, who had worked in that position at several medium-sized companies, got much of his capital from a Massachusetts firm, Analog Devices, which was eager to get into digital signal processing (DSP). (An example of DSP is collecting temperature readings from an industrial furnace, converting them to digital form, processing them by computer, and returning them in analog form to control the furnace.)[12]

Analog Devices hoped that its use of BIT processes would result in new customers in scientific, military, and medical institutions. In return for Analog's investment (which came through its venture-capital arm, Analog Devices Enterprises), BIT agreed to make its technology available to the company, which acquired about 40 percent of the new firm. Analog's investment gave BIT financial respectability. Other venture capitalists, including First Interstate Capital, BancBoston Ventures, Raytheon Company, Union Venture Corp., and InterVen Partners, followed its lead and obtained about 20 percent of the equity. The state of Oregon also joined in with an industrial-revenue bond that financed the construction of BIT's manufacturing plant.

In April 1986 BIT announced that its product was ready to market. The product, called the B3018 chip, had five times the speed of other ECL chips on the market and ten times the power-use advantage over traditional ECL processes. The company hoped to reach a specialized market (as opposed to a mass one like automobiles) with its new device, but the competition promised to be fierce, not only from GaAs producers but from several companies that planned to bring improved, high-performance complementary-metal-oxide-semiconductor chips (CMOs) to market.[13]

Another entry in the minisupercomputer field came in 1983. Amidst great controversy and publicity, Sequel (soon to be called Sequent) Computer Systems was created by seventeen former employees of Intel, led by Casey Powell, who had been general manager of the microprocessing unit there. It was not a friendly spin-off. Intel was so shaken by this large-scale departure that it threatened legal action against those who were leaving, although it did not follow through with its threat.[14]

What attracted outside attention was the large amount of venture capital ($5.2 million) raised by the new company in March 1983. Also surprising was the prestige of the investing firms, which included some of the nation's best—Kleiner, Perkins, Caufield and Byers; Venrock Associates; and Hambrecht and Quist. Institutional Venture Partners of Menlo Park, California, the lead firm, and Shaw Venture Partners of Portland (underwritten by U.S. Bancorp) were also among the original investors. So appealing were the prospects of the company that twice as much money could have been raised. One of the investing firm's officers said that Sequent had the potential to be a "major, major" firm in the Pacific Northwest, which would "do a lot for the Northwest and will make their investors a lot of money."[15]

Eighteen months after the initial announcement, Sequent announced its first product, the Balance 8000 computer, a 32-bit-processing machine. The announcement was accompanied by even more publicity than that hailing the firm's formation. Some observers predicted that Sequent would compete with Mentor Graphics to become the next Oregon firm on *Fortune*'s list of America's five hundred largest companies. One of the investors foresaw that Sequent might well create another VAX, a product of Digital Equipment Corporation that had set the standard for minicomputers, and the company would be as attractive an investment as the lucrative firms of Apple, Apollo, and Intel.

Right on schedule, in December 1984, Sequent shipped its first machine. During its first year the company recorded $1 million in sales.[16] In the summer of 1985, it signed an agreement for its first major shipment—a $10 million order—with Teradyne of Boston, a leading manufacturer of automatic test equipment. For several months before announcing the agreement, the two firms had been engaged in joint development work. Sequent's Balance 8000 machine was to be integrated with Teradyne products and sold to end users. The Sequent product would enhance the productivity of the Teradyne machines. The year 1985 was also marked by another round of venture-capital funding and the introduction of a second, more powerful machine, the Balance 2100. One of its first purchasers was Purdue University.[17]

Sequent's success story continued into the next year. In March 1986 the firm announced a contract with the $20 billion German electronics firm, Siemens AG. The amount of the order—to be shipped over five years—was $50 million, the largest ever written in the parallel processing field. The agreement also involved technological exchanges between the two companies and a licensing agreement to allow the German firm to produce some system parts on its own. For Siemens, the decision to buy from Sequent saved the company about two years' time in developing its own parallel-processing machines. For Sequent, the income and prestige attached to the sale were of enormous value, but the arrangement also made parallel processing much more attractive to competitors in an already competitive world.[18]

Parallel processing was thus not the only technology available for supercomputing. There were other approaches to the minisupercomputer market, which was estimated to grow from $125 million in 1986 to $1 billion in 1990. In reaching for this bonanza, there was a danger that the minisupercomputer market would repeat the history of minicomputers and personal computers. In those industries the struggle for lucrative markets resulted in the demise of most competitors, because there were too many companies pursuing too few customers. Equally troubling to minisupercomputer officers, however, were the numerous approaches to parallel processing, "each with its own claims to price and performance, [which] led to intense confusion by potential customers." Competition has caused slower market growth than had been anticipated, a condition that further threatens the staying power of the competing firms.[19]

PART 2

THE LARGER COMMUNITY

7

High Technology and Education

THE LOFTING OF THE SOVIET SPUTNIK in the fall of 1957 created a crisis of confidence in the United States. Politicians, newspaper and television commentators, military personnel, and ordinary citizens all became terrified that the Russians had attained scientific and technical supremacy over America. Amidst the outcry over what to do, President Dwight D. Eisenhower and Congress fixed on both short- and long-term programs to get the country moving in the space race. Among them was the epochal National Defense Education Act of 1958, the largest program of federal aid to education in American history, creating a system of loans and fellowships for undergraduate and graduate students and a program of grants for the construction of university buildings. In 1962, it was followed by a second law that added to the federal largess. Science and technology were the major beneficiaries of the act.

While the two measures had great influence across the United States, their benefits were not equal and universal. One region partially left out was the Portland metropolitan area. Portland's problem was that it had no large public university to benefit from federal aid to education. In the city were Reed College and the University of Portland, both of which were private, almost exclusively undergraduate institutions, and Portland State College, a public institution derived from a postwar veterans' extension center that had been given four-year college status in 1955 over the opposition of the University of Oregon and Oregon State College. After the furor over sputnik and the passage of national education laws, Portland's scholastic deficiencies came to the attention of several influential people, who realized that it was the biggest city west of the Mississippi River without a large university.

The first to express his concern was Oregon's governor, Mark Hatfield, who had made economic development one of the features of his administration since his election in 1958. Among other things, he had spent months trying to persuade new industries to come to Oregon by offering them a variety of incentives. A former political science professor at Willamette University, the governor wanted to strengthen the state's economic base by urging greater efforts to promote science and technology in colleges and universities, particularly in the Portland area. In 1959 he called for a report from a committee on science, engineering, and new technologies, a part of the Oregon State Department of Planning and Development. The committee, chaired by Walter P. Dyke, a physics professor at Linfield College, reported in May 1961 in favor of a new, independent graduate institution in Portland that would provide educational opportunities for students and, significantly, research to "serve all resident industry as well as new and prospectively new science-based industry."[1]

Hatfield was pleased with the report and appointed an eight-person Advisory Committee to pursue the matter of higher education facilities in Portland. Richard H. Sullivan, president of Reed College, chaired the committee of which Howard Vollum was a member. In July 1962 the group reported in favor of an Oregon Cooperative Center for Graduate Education and Advanced Research. The main point made in the report was that a cooperative graduate center should be formed from existing public and private universities. It specifically rejected the creation of a new institution.[2]

The committee recognized the need for a new building and for a faculty drawn from existing universities and its own members. It recommended that the center be situated on the west side of Portland, mentioning that area's proximity to the Oregon Regional Primate

Center and the new electronics firms. The center would not grant its own degrees, but offer its facilities to students and faculty of existing public and private institutions. It was to be financed largely by private donations. The initial course offerings would be electronics (mentioned first), biochemistry and health-related sciences, metallurgy, and wood-products chemistry. The purpose of choosing such fields was that "these seem most likely to build bridges most rapidly between mutually supporting efforts of the Center and of industry and thus to be more closely related to the end of economic growth and development of Oregon."[3]

The report clearly recognized the need of electronics industries for graduate education in Portland, the rivalry among state institutions and between them and private institutions, and the difficulty of raising state funds for higher education. It thus chose a carefully balanced approach that it hoped would be a beginning for advanced training in science and engineering in the metropolitan region. However, it did not result in action.

In January 1963 Hatfield appointed a third group to attempt to implement the report's recommendations. Meanwhile, in their 1963 regular session, state legislators had adopted a resolution introduced by Sen. Don S. Willner of Portland calling upon the State Board of Higher Education to "establish a quality program of graduate education in the arts and sciences in the Portland metropolitan area." One of the main justifications for the program was that it would "help attract new science-based industry and payrolls to our state and [would] make an essential contribution to the development of current business and industry." Again, the resolution did not call for either a new university or university status for Portland State College, but left the situation fluid.[4]

Many influential Portlanders were disappointed by the reports, believing that they neglected the needs of the Portland area (which in 1960 contained 41 percent of the state's population). They also believed the reports represented the governor's bias against public institutions and the hostility of existing colleges and universities to the idea of creating a new university in Portland.

Some of the dissenters soon moved to open another avenue for strengthening scientific and technical study and research by proposing that the Portland City Club study the problem again. The club agreed and charged a committee under the chairmanship of Robert H. Tarr, president of the Northwest Science Investment Corporation, with studying and reporting on "the feasibility and desirability of establishing facilities for graduate study and basic and applied research in the Portland metropolitan area" and recommending "an action program for realization of such facilities as prove feasible and desirable." The committee's members included Douglas Strain of Electro Scientific Industries and Samuel L. Diack, a physician and vice president of the Oregon Medical Research Foundation, who was deeply committed to scientific research. The committee reported on 1 April 1963.[5]

Its conclusions were threefold. It recommended that the State Board of Higher Education transform Portland State College into a university as soon as possible; that "there be established in the Portland Metropolitan area a privately supported center for advanced study and research at the post-doctoral level to further integrate physical sciences, engineering, and mathematics with the life sciences—medicine, biology, and medical psychology"; and that a privately funded industrial-research institute be created.[6] The report contained something for almost everybody, but its full implementation would require support from a variety of often clashing constituencies.

The result of all these reports and aspirations was not the development of Portland State College into a major university. The opposition of the University of Oregon and Oregon State University and their numerous and influential alumni made the development politically impossible. Instead, a private institution, and a most unusual one, the Oregon Graduate Center, was developed.

THE OREGON GRADUATE CENTER

The Oregon Graduate Center traces its roots to 24 January 1963, when Governor Hatfield created his fifteen-member board to put the Science, Engineering, and New Technologies report into action. For the most part the work involved in founding the center was carried out by Howard Vollum of Tektronix, John Gray of Omark Industries (a manufacturer of sawchains), Ira Keller of Western Kraft (a papermaker), and Samuel Diack. These men made the initial financial contributions and drew up a charter, approved by the State of Oregon on 2 April 1963, that stated the purposes of the center, including an industry-related aspect that has always remained, though it has not always been emphasized. In planning the institution, the founders relied in part on the advice of consultants, one of whom was Donald Benedict of the Stanford Research Institute. Among the things he was asked to do was to draw up a job description for the center's president. In time, the founders came to believe that Benedict himself best fitted the description, and they hired him as center's first president in September 1966.[7]

As President Benedict began fleshing out the charter and hiring the first faculty, he was working against four developments that influenced his presidency. One was the desire of industry (and not simply high technology) for applied research to aid in its growth. Another was the enormous expenditure of federal funds to assist science and engineering after sputnik. A third, more an attitude than a development, was the traditional belief of American professors and graduate students that theoretical ("pure") research was far more valuable and certainly more prestigious than applied research or engineering. The president (and the trustees) were also influenced by the models for the center, especially Rockefeller University in New York City.[8]

Oddly, Benedict chose for the faculty scientists who were dedicated to, and intended to remain dedicated to, "pure" science. Although they were promising and came from excellent institutions, their selection, and the selection of students with similar attitudes, seemed to deviate from the charter. They helped give the center a sort of "ivory tower" reputation that shadowed the Benedict era and later took some time to redress.

Two other features of the organizing years that distinguished the graduate center from traditional educational institutions were the absence of departments (though this was later changed) and of tenure.

The lack of departments reflected the current interest in interdisciplinary work and a belief that academic departments impeded it. The lack of a tenure system probably came from the belief that it would be inappropriate to make long-term financial commitments in those early days and from the business-dominated board of trustees, who saw tenure as a refuge for "nonproducers." The original institutional structure, however, did include a faculty committee system and a faculty senate, which ensured faculty participation in the institution's operation. Indeed, Benedict often deferred to the faculty, some of whom remembered him as being "afraid" of them.[9]

The Oregon Graduate Center's first campus was at 9430 SW Barnes Road in Washington County, adjacent to the Sunset plant of Tektronix on land that had been vacated by the Martin-Marietta Company. Tektronix had persuaded Martin-Marietta to sell the land at a nominal price, and the Tektronix Foundation had bought it and in turn donated it to the center.[10] In 1967 the first faculty members were hired, and in the fall of 1969 the first students took up residence. If newspaper coverage and support are any measure, the public was enthusiastic about the center's prospects. Yet these promising beginnings did not bear fruit immediately, and the Benedict years saw many disappointments.[11]

For one thing, the president and the board did not make a successful start on solving the problem that has been the center's bane from its beginning: finances. In keeping with his "ivory-tower" orientation, Benedict decided to discourage faculty members from raising support through federal and industrial grants, assuming that the board would raise their salaries and expenses and furnish funding for the students. This soon proved to be impractical, and the center's existence became dependent on periodic contributions from a few individuals, mainly Vollum. The president was also unable to obtain state funding. The center did get Hatfield to recommend an appropriation of $1.5 million (later reduced to $596,000) for the 1965-67 biennium, but the legislature refused to allocate any money to the fledgling institution. Benedict also erred in not attempting to hire, among his first faculty, scientists with established reputations who could gain immediate recognition of the institution.[12]

Benedict's major accomplishment was in securing a seventy-four-acre permanent campus site on Walker Road, the Donovan farm, from the Tektronix Foundation. There, on 15 August 1969, the first new building was dedicated. However, the building's construction was plagued by financial misjudgments on the part of the president and the board, who underestimated the costs of scientific research and instruction. As a result, plans were changed two or three times during the construction process, multiplying the cost considerably. Twelve days after the building was dedicated, the trustees dismissed Benedict.[13]

The Graduate Center subsequently reached its nadir, hiring nineteen faculty members for the 1971-72 academic year (earlier the figure had been as low as twelve). However, there were only five students in 1969 and 1970, and by the summer of 1972 they had been told to go elsewhere to finish their education and further their careers. All the faculty members were seeking other positions.

Benedict was followed by officers who were interim, or ineffectual, or both. One of the presidents wanted the institution to become a part of the Oregon State System of Higher Education, but the trustees feared that the change would negate any chance of obtaining scientific research of industrial value. (It would also require the introduction of a tenure system.) There was also some talk of annexing the center to Lewis and Clark College, or even of Tektronix's buying it.

Rather than giving up, the faculty took the lead in saving the institution. Nick Eror, Richard Elliott, David Roe, and a student member formed a committee that devised a plan for the faculty to raise 80 percent of the Center's income. A committee of the board was formed, but failed to act. In frustration, another faculty delegation went to trustee Ira Keller.[14] Keller not only took an interest in the committee, but board members took an interest in him to fill the leadership void. In the summer of 1972, they appointed him president.

The new president was a man of many parts. With an undergraduate degree in engineering from Rensselaer Polytechnic in 1920, he had gone into the paper business and risen to be executive vice-president of the Container Corporation of America. In 1954 he founded the Western Kraft Company (now Willamette Industries) in Oregon. While at that company, he had been chosen to head the Portland Development Commission (from 1958 to 1972) and had supervised the city's urban-renewal program that had erected the South Park Blocks Project. A man of controversy throughout his career, decisive if not dictatorial, Keller had enormous enthusiasm for whatever task engaged his attention. In his lifetime he was somewhat larger than life; after his passing, he came to be regarded by some as the savior of the Oregon Graduate Center. Although this estimate may be an exaggeration, he made many contributions to the center, the most important of which was releasing the talents of faculty members, who later paid a price for their new freedom.[15]

During the Keller years, the president encouraged faculty members to pursue their own grant funding and worked with them to restore the original charter's intent of concentrating on applied research. For example, through his connections in the paper industry, he was instrumental in persuading the Weyerhaeuser Corporation to fund an experimental tree-cloning project at the center.

His connections in the East and Midwest were also helpful, particularly in publicity. An article he secured in *Business Week* extolled the Center as "an observer lab [that] does big jobs for industry." He persuaded his corporate acquaintances in Portland to make donations and secured a commitment from board members to contribute.

Where the faculty was concerned, Keller set up four departments and recruited specialists as department heads to carry on research projects of interest to local industry. In 1974 Fred Holmes, Frank Hauser, and William Woods were hired in electrical engineering, chemistry, and materials science, respectively, and in 1978 Gail Massey came as a laser expert. Even the new vice president, Richard Kerr, was a specialist, and the older faculty began to shift their focus from theoretical to applied research.[16]

Once Keller had hired new department heads, he instituted a controversial new policy of letting them dismiss "surplus" faculty. (Subsequently, it became clear that not all the dismissals were being made on objective grounds.) He also set up an administrative committee structure under his control that paralleled the faculty committees and used his structure to bypass the faculty system.[17] Such high-handedness did not sit well with all faculty. Some resigned, while others waited for proof that his changes were indeed wise.

At the time, students were none too numerous and in 1975 no degrees were granted. Yet there were small signs that the Oregon Graduate Center was active. In 1973 the first doctorate had been granted, and in the same year the school was fully accredited (with fourteen students it was the smallest accredited institution in Oregon). Then, too, grants began to arrive. In 1973 the National Science Foundation gave the center $50,000 for a laser-research project headed by Massey. A year later the Hill Family Foundation gave $55,000 to establish the Northwest Air Pollution Laboratory. Later the same year, the center attracted much public attention when it received the largest outsider research grant ever given by the Weyerhaeuser Corporation. The $1.2 million gift was intended to support a five-year project headed by G. Doyle Daves, professor of chemistry, to produce trees from living cells rather than from seeds. In 1974 the center faculty also received a grant to design an electronic shutter that would give laser pulses a precise shape needed to heat fuel pellets for nuclear fusion research. And two other grants—from the Army and the Air Force, respectively—allowed faculty to investigate both the transmission of laser beams through the atmosphere and the use of lasers to measure wind velocity.[18]

Keller retired from the presidency in 1977 and assumed the position of chairman of the board. During his last months as president, he had been grooming Richard Kerr to succeed him. The latter's tenure as president from 1977 to 1979 was a stormy one. His personal and ideological conflicts with faculty and members of the community culminated in a successful move by the department heads and some administrators to force him out of office. After a faculty vote of no confidence, he resigned. Despite the turmoil, his presidency was marked by some notable achievements, among which were a $1 million grant from the M.J. Murdock Charitable Trust for the construction of a new wing in the campus quadrangle. Another was the appointment of John S. Blakemore as professor of applied physics and electronics. An expert in semiconductors, Blakemore was the first faculty member appointed in solid-state physics, the foundation of the electronics industry. A third development in 1978 was the donation by Omark Industries of the most powerful laser possessed by an educational institution in the Pacific Northwest.[19]

Kerr's successor was F. Paul Carlson, who was appointed president in 1980. With a doctorate in electrical engineering from the University of Washington, he had come to the graduate center in 1977 as vice president for industrial affairs and development and as professor of applied physics and electronic sciences. In the year since Kerr's resignation, he had also served as acting president.

Even before he became president, Carlson was the moving spirit behind the Oregon Graduate Center Science Park, a major change in direction for the center that promised to end its financial woes and to foster the economic development of Washington County.[20]

Since 1951, when Stanford University had founded the first science park, the idea of clustering industries adjacent to universities had been widely discussed in American high-technology circles. Some of the Oregon Graduate Center's founders—Vollum, Diack, and Walter Dyke—had envisioned a science park and had commissioned a physical model of one. During the seventies, the idea continued to be discussed in a desultory way but was realized only after Carlson became president.

Carlson saw a science park as a means of giving the Oregon Graduate Center the sound endowment that it had always lacked and of more effectively demonstrating the center's mission of close industrial-university cooperation in teaching and research.[21] In May 1982, the center announced its plans for the science park, which was to be constructed on one hundred sixty acres of land that it owned next to its current site. The $120 million project was to be leased to the Rembold Corporation, which would undertake the building, make site improvements such as roads and sewers, and lease the property to tenants. It was hoped that high-technology firms would thereby gain attractive sites as well as the educational advantages to their employees of working with research scientists at the center. A holding corporation, the Oregon Graduate Center Corporation, was created to oversee both the center and the science park. Early publicity asserted that the park would provide employment for four or five thousand people. Predictions were that it would be completed in five years and would produce, in endowment, an annual income equivalent to $4 million. Carlson frequently described the park as the Oregon Graduate Center's endowment and claimed that it represented the first step in a decade of rising employment in the Sunset Corridor which by 1993 would reach thirty-eight thousand new jobs in high technology and ninety-five thousand more in support services.[22]

In July 1982 the center and the Rembold Corporation sponsored "Reaching for Oregon's Horizon," a conference that was attended by business and political leaders and served to launch the park in a well-publicized manner. The conference promoted the idea that, besides fostering industrial and academic relationships in developing products and teaching employees, the park would help "a broader segment of society in understanding the implications of scientific and technological innovation." The center planned to invite "guest humanists and other life scientists working in history, literature, and politics, to take up residencies... [to] allow a fuller exploration of science, its human implications and a bridging of the cultural gap." The park was also to help improve the science education of high-school students and their teachers and furnish community-college and continuing-education work. If these ambitious objectives were not all as novel or as precise as their proponents maintained, they were at least well intended. To realize them, however, the center first had to liquidate its $1 million land debt, get the necessary legal underbrush cleared away to ensure that it would not lose its tax-exempt status, and construct and occupy the first buildings.[23]

Carlson's enthusiasm, ingenuity, and charisma enabled the center to meet these financial and governmental objectives. By August 1983 the first tenant, Planar Systems, had been secured and ground for the seventy-thousand square-foot-building it would occupy had been broken. In the enthusiasm of the moment, the center's officials envisioned wider ranging visits by Oregon Graduate Center faculty to attract more and more companies to the park—"like filings to a magnet," in the words of one of them. This, in turn, would make the Oregon Graduate Center a Pacific Northwest version of the Massachusetts Institute of Technology, and the Sunset Corridor would become the western equivalent of Boston's Route 128.

Planar moved into its facility in February 1984. By April 1986 nine other tenants had followed. As the science park expanded it required full-time leadership, and in August 1985 Carlson was released from the presidency to assume that of the Oregon Graduate Center Corporation. When Rembold withdrew from its roles as developer and landlord, the science park also had to assume responsibility for those functions. Whether the Oregon Graduate Center Science Park will fulfill its many roles is, like so many aspects of Washington County's high-technology industry, open to conjecture.[24]

Besides creating the science park, Carlson also took pride in expanding the center's ties to the local high-technology industry through the creation of a department of computer science and engineering which came into existence in 1981. With financial support from Tektronix, Intel, and other corporations, foundations, and individuals, a computer-science laboratory was constructed and faculty recruited. However, this was not accomplished without incurring the resentment of some faculty members in traditional departments, who believed that it would have been advisable to strengthen their areas before undertaking expensive new ventures.[25]

Carlson was also successful in publicizing the center in the larger community through other new directions, such as the Saturday Academy for gifted elementary and secondary students and a quarterly magazine, *Visions*. From 1980 to 1986, he increased the number of students (41 to 103) and faculty (24 to 43).[25]

By 1986 the Oregon Graduate Center was a distinctive institution. Although it resembled traditional universities in having as its purposes teaching and research, in its departmental structure (applied physics and electrical engineering; chemical, biological, and environmental sciences; computer science and engineering; and materials science and engineering) and in its academic ranks, it also differed from those institutions. Like the original Johns Hopkins University (founded in 1876), the center has no undergraduate students. Its full-time student body remains small, numbering 105 in the 1986-87 academic year. The teacher-student ratio of 2.5 to 1 is one of the most favorable in American university life and in principle produces more of a collegial than a mentor-pupil relationship.[26] In almost all aspects of student and faculty work, the center is oriented to the needs of industry—an orientation that has proved worthwhile in several ways. Some of its current projects of industrial and community value range across the spectrum of the mathematical, physical, and biological sciences. For instance, Floating Point and other companies producing software and hardware for supercomputers rely on parallel processing to solve complex problems rapidly. At the center faculty and students are working on parallel processing from three directions: developing a new computer language in which programs are mathematical functions; devising new programming methods; and investigating new designs for parallel-processing computers.

One of the negative aspects of high technology is the actual or potential poisoning of groundwater by escaping hazardous wastes, as we shall see in chapter 10. At the center's Water Research Laboratory, researchers are studying the movement of chemicals through rock and soil in eastern Oregon, California, and Canada, at sites where hazardous waste has been placed. Scientists are also studying gallium arsenide and other semiconductor compounds used in electronics. In other realms, center researchers are working on cloning trees and developing tissue cultures that may produce "super trees," while others are investigating such problems as improving the life span of railroad tracks to carry increasingly heavy loads at increasingly fast speeds.[27]

Less constructively, the faculty and the larger community are denied the benefits of academic freedom and job security that accompany the tenure system. The institution has little or no endowment, although it is hoped that the science park will correct this situation. So far, limited endowment funds and sporadic financial contributions by trustees have caused the faculty to raise most of its own salary.

Notwithstanding all these difficulties, the faculty has persisted and endured. Not only endured, but flourished. The center's faculty members have won several awards and received many distinctions. It is to them that most of the credit for this unusual institution is due.

STATE FUNDING FOR HIGH-TECHNOLOGY EDUCATION: THE QUEST

Everyone who dreams of a flourishing high-technology community realizes the importance of higher-education programs and facilities to furnish a supply of undergraduate and graduate engineers, computer scientists, and managers who understand technology; to retain them in their later careers; and to provide research facilities for advances in knowledge. For many years the Oregon Graduate Center was the only specialized institution in the state dedicated to high technology, and it was not until the early eighties, as high technology came to be regarded as the salvation of Oregon's economy, that technical and scientific education began to receive more public attention. With this attention came hopes of increased state funding, and these hopes in turn led to a struggle to determine which institutions would receive public funds and which educational philosophy would triumph.

On one side in this contest was the Oregon Council of the American Electronics Association (AEA), the Oregon Educational Coordinating Commission, the Oregon Graduate Center, many officers of high-technology corporations, and legislators from Washington County and elsewhere, who wanted an immediate supply of engineers for the benefit of high-technology firms. The goal of this coalition may be summarized as practical research of the type that has characterized the Oregon Graduate Center in its later years, while it was determined that high-technology educational facilities should be conveniently situated for the business firms of Washington County.

On the other side was the Oregon State System of Higher Education, especially Portland State University, certain legislators from Portland, and some high-technology executives from the Portland area. This group wanted to extend the engineering offerings at Portland State University, basing its argument on the principle that—at least at the undergraduate level—an engineering education must not be exclusively technical, but should include a fair range of liberal arts courses. Proponents also believed it wiser to develop an existing public university, in the largest city in the state, than to infuse money into a private school or to create a new hybrid institution in Washington County. Complicating the contest was the historic opposition of the University of Oregon and Oregon State University to the development of Portland State University, a position that made it difficult to present to the opposing side a united state system front. Out of these two sets of opponents arose the tortuous compromises reflected in high-technology public education in the first half of the eighties.

The maneuvering over who was to be the educational beneficiary of the high-technology phenomenon began in the special session of the 1982 legislature. The session met to find a way out of the state's budgetary problems, caused by a $314 million shortfall in revenue projections. In these inauspicious circumstances, Rep. Vera Katz, a Portland Democrat and an employee of Portland Community College, devised a response to the wishes of the high-technology community. Working with the Oregon American Electronics Association council, she guided through the legislature a bill establishing the Oregon Consortium for High Technology Education. The state was to contribute five hundred thousand dollars to the consortium, which would be matched by private donors. After presenting the plan to the legislature, Katz was asked by Governor Victor Atiyeh to make her plan a part of his economic recovery program. She agreed, thus cementing formally the link between education, high technology, and economic progress.[28]

Katz also insisted that the moneys be allocated by the Educational

Coordinating Commission rather than the board of the State System of Higher Education. She justified this arrangement on the grounds that she wished to include private and community colleges among the recipients and that the commission had coordinating authority over all Oregon's educational institutions, private and public. She also argued that if the state system disbursed the money, it would be distributed equally among the state institutions rather than to exceptionally promising programs.[29]

The details of the plan were worked out by Katz, the governor, and the Educational Coordinating Commission without meaningful representation from the Oregon State System of Higher Education. The plan passed the legislature, and the governor appointed the members of the consortium:

> Gene Chao, president of Metheus Corporation, representing small high tech firms; Paul Lintner, president of Electro Scientific Industries Inc., representing medium sized high tech firms; Philip R. Robinson of Tektronix representing large sized high tech firms; Kenneth Moyle, general manager of Intel Corporation, representing local subsidiaries of large high tech firms; John Anthony, president of Portland Community College, representing the community colleges; Paul Carlson, president of Oregon Graduate Center, representing private education institutions; Louis Perry, president of Standard Insurance Company, representing the Board of Higher Education; and Ross Thompson, chairman of the Educational Coordinating Commission, representing that body. Senator Jim Gardner, D-Portland, and Katz served as ad hoc members.[30]

It should be noted that no institutional president from the Oregon State System of Higher Education was appointed.

Significantly, the consortium selected electronics engineering and computer science as the fields on which it would focus. In its fund-raising and granting programs it concentrated on the Portland area, although it was careful to announce as its ultimate aim the development of a "Willamette Valley High Technology Corridor."

By January 1983 matching funds had been secured and the first grant recipients had been selected. The grant program advanced in two phases, in 1982 and in 1984-85, respectively. By January 1986 most of the money for both phases had been awarded. The total sum expended was approximately $3 million, of which the state and industry each contributed half. The recipients numbered six community colleges, one private college, two private universities, and three public universities. Funds were expended on thirty-two projects, examples of which included intensive training for Umpqua Community College staff in the use of the college's new electronic drafting and computer-assisted design and manufacturing equipment; a new faculty position in solid-state physics at the Oregon Graduate Center; and electrical engineering and research equipment, and the computer vision and laser-system laboratories, at Portland State University.[31]

Since the Oregon State System of Higher Education had been given a relatively small administrative role in the consortium, the new chancellor of the system, William E. "Bud" Davis, found support to form the Council for Advanced Science and Engineering Education/Research for Industry. Although the public universities had given engineering courses for almost a century (the first engineering school began at Oregon State in 1893), not enough programs measured up to the standards of the high-technology industry. In July 1982, shortly after he was hired, Davis presented a "Proposal for a High Technology Engineering/Computer Science Educational Program for Industry." The document was based squarely on the need for technological education to help solve the state's economic problems, and it opened with a quotation from the report of the governor's education task force of two months earlier: "Oregon must support higher education, both public and private, with a moral and physical commitment to excellence if the state is to compete with the rest of the nation culturally and economically." In the report the chancellor went on to make several points about the relationship of higher education and high technology that had been made twenty years earlier, but with the exception of the Oregon Graduate Center, had produced few results. The time demanded their restatement.[32]

High technology bulked large in the report. A statement on the first page underscored the point: "Oregon has many attractive features that make the development of existing industry and the attraction of new industry, *particularly high-technology industry*, possible." It predicted a "substantial" concentration of high technology in the state by the year 2000, particularly in the Willamette Valley. The report maintained that public and private institutions could contribute to this development by identifying the needs of industry and matching them with university resources; by "strengthening academic and research programs in areas essential to high-technology industry"; and by developing "coordinated and effective educational programs to serve the needs of Oregon and its industries."

For his part, the chancellor promised to appoint members to the council from representatives of the Oregon State System of Higher Education institutions, the private universities, and the high-technology industry, headed by a director responsible to the chancellor. The council would be charged with carrying out the three programs described above. The proposal argued for an Oregon State System of Higher Education program in electrical and computer engineering and computer science, based at Portland State University, which would operate in conjunction with Oregon State University and the University of Oregon.[33] It was significant that, in introducing his proposal, the chancellor was careful to allude to his being guided by proposals of the Oregon Council of the American Electronics Association and the Educational Coordinating Commission. The heart of the proposal was a request for nine full-time faculty positions in Portland, to be filled from the three major state universities, although PSU was to "play a central role." The total instructional budget request was $1 million for the 1982-83 academic year. This was regarded as "seed money" for the program, the term advisedly implying that further moneys would be needed in the future. The report concluded with "institutional profiles" of the three state universities, which included detailed information about their degree programs and areas of graduate specialization in computer science and electronics engineering relevant to the needs of the high-technology industry.[34]

Linkage between the Oregon State System of Higher Education and hopes for state economic development continued. At a meeting in August 1982 of the Education Commission of the States in Portland, the Oregon State System of Higher Education and Governor Atiyeh jointly sponsored a presentation by the governor of North Carolina and the president of the North Carolina System of Community Colleges. In his address, Gov. James Hunt stressed the reciprocal dependence of economic development and educational improvement, declaring: "The economic future of this nation is with high technology, and that is going to require close and continuing cooperation among the universities, community colleges, the executive and legislative branches of government, and the private sector." In 1982 Oregon State System of Higher Education also began to prepare a strategic plan for the system that began with a review of the strengths and weaknesses of existing programs and how they related to economic development.[35]

As the High Technology Consortium and Council for Advanced Science and Engineering Education/Research for Industry took form in 1982 and 1983, interest in the high-technology relationship continued. For example, in April 1983 a group of state legislators toured six Washington County high-technology industries and the Oregon Graduate Center and were told repeatedly, in the words of one corporate executive: "We need a good education system that can provide continuing education for engineers and computer scientists to enable us to attract the kind of people we need. Education is the backbone of our industry." Incentive to keep the education issue alive was fostered by the competitive situation of Oregon's high-technology firms. As the depression of the early eighties continued, practically every state in the Union was desperately attempting to attract these new "wonder industries."[36] Practically every state thus had to promise increased expenditures for education, especially advanced engineering and scientific programs.

This competition hit home with news of Washington State's education budget for the 1983-85 biennium. While Oregon was spending $500,000 of public funds on the High Technology Consortium, the

Washington legislature designated $10 million for education programs designed to meet the needs of high technology and created a technology research center funded at between $12 million and $14 million to pool the resources of public universities and private industry. Among other political leaders in Oregon, Atiyeh was convinced that higher education needed new state funding to meet the competition. In a portion of his "State of the State" address on 7 March 1985, he recommended the creation at the state-supported universities of three "centers of excellence," all of which would have some bearing on high technology: an international trade and commerce institute at Portland State; an institute for electrical and computer engineering at Oregon State; and an institute for biological and advanced sciences at the University of Oregon.[37]

By 1985 it had become evident to many citizens that neither the High Technology Consortium nor the Council for Advanced Science and Engineering Education/Research for Industry had fulfilled their proponents' expectations. The American Electronics Association Council in Oregon and a companion organization, the Business-Education Compact of Washington County, then took a new tack. They were able to anticipate additional funding for their next proposal because the voters had authorized a state lottery in the fall election in 1984. In the legislature's 1985 session, the American Electronics Association was able to obtain funding for the creation of the Oregon Center for Advanced Technology Education, another in the long series of proposed university-industry cooperative ventures.[38]

Again Katz and the high technology lobbyists were at the forefront of the new creation. However, the legislators were more specific about funding the institution than about its mission. They allotted $1 million in lottery funds (subsequently reduced to $500,000 because of a shortfall) and the balance of the moneys from the Council for Advanced Science and Engineering Education/Research for Industry to the new institution. At the same time they left its precise goals (to avoid interinstitutional conflicts while the legislation was being written) to the new Governor's Commission on Technical Education created on 3 June 1985 with Earl Wantland of Tektronix as chairman and Tom Bruggere of Mentor Graphics as vice-chairman. The group would also serve as an advisory body to the new center. (In addition, an Advisory Committee to the Commission on Technical Education was appointed, comprising public and private university and college presidents and chaired by President Larry J. Blake of the Oregon Institute of Technology.) Of the remaining eight Technical Education Commission members, five were presidents of Washington County firms. The group's statement of mission proclaimed:

> High technology industry and its employees, the higher education system, both public and private, and the State of Oregon and the forces of economic development all benefit from the development of programs in Washington County. . . . It offers for the State of Oregon, for a relatively small investment of financial resources, the opportunity to leverage its valuable but limited higher education resources in a program that will have long-term benefits to the state's economic development goals. Those benefits include keeping Oregon's existing advanced technology community strong and helping to attract new high technology companies to the area because of the high quality educational programs designed specifically for such companies. Through the development of a state-wide televideo link, other communities, advanced technology companies and higher education institutions in our state, including community colleges, also could benefit from OCATE offerings. In short, OCATE can offer substantial incentives for all its constituencies.

It also stated: "Programs *initially will be aimed at the high technology community in Washington County* [italics in original]." The new center was to be administered by the Oregon State System of Higher Education, and it was to be headed, at the request of industry, by a special vice-chancellor.[39]

The Oregon Center for Advanced Technology Education's first goal was to provide master's and doctoral programs for the high-technology community. It was also given responsibility for a master's degree in business administration, sponsored jointly by the University of Oregon, Oregon State University, and Portland State University. The faculty for the programs was drawn from the state's higher-education system schools, private schools such as the Oregon Graduate Center, and visiting faculty from outside the state. The site of these activities was the Rock Creek Campus of Portland Community College. The institution was put together with great speed, and the first classes (six in all) were taught in the winter term of 1986. In addition, there were three conferences in 1986: one on artificial intelligence, one on electroluminescence, and one in cooperation with the American Vacuum Society on silicon materials.

The structure, curriculum, and mission of the new center were again compromises, giving something to the Oregon State System of Higher Education, Oregon Graduate Center, Portland Community College, and most of all Washington County's high-technology firms. The Oregon Graduate Center gained an entering wedge for the state funding it had long desired, while the state system had the prestige of administering what was conceived as a glamour institution.

Omitted from the Oregon Center for Advanced Technology Education was the area of the state outside Washington County. When the chancellor first proposed the new center to the Oregon State System of Higher Education, some members of the board voiced the fear that it was too narrowly focused geographically. The chancellor told them that the Council for Advanced Science and Engineering Education/Research for Industry would continue and would provide the economic development needed in what was popularly becoming known as "the other Oregon." Another concern was, in one member's words, "the dilution of science [Oregon State System of Higher Education] resources to provide, in effect, a private graduate school for one industry." The chancellor assured him that he would work for additional continuing-education facilities in other areas of the state.[40]

COOPERATING WITH THE COMMUNITY

High technology has also affected education at the community college, secondary, and elementary levels. The most ambitious effort in this direction was made by Gary Conkling, a Tektronix executive who conceived of what became known as the Business-Education Compact. In the spring of 1983, Conkling and others in the Beaverton and the Hillsboro chambers of commerce were considering the question of economic development. They decided that all the area's developing business firms—at least those in high technology—were fed by education and that this dependence should be formalized rather than proceeding along separate paths. The group also decided to expand beyond Beaverton and Hillsboro to include Forest Grove, Tigard, and Tualatin.

Early in its deliberations, the planning group decided to study high-technology areas and models of cooperation between industry and education in Dallas, Boston, Denver, San Diego, and Seattle.[41] From these models and members' own experience, the business committee conceived of a task force made up of representatives of all types of educational institutions, from the Oregon Museum of Science and Industry to elementary schools. The final task force numbered thirty-two, half business and half education from kindergarten through graduate school.

The group chose a small number of projects for attention during its first year of operation. The first, launched in January 1985, involved providing a postsecondary high-technology center in Washington County for students not choosing to pursue a four-year college or university education. John Anthony, president of Portland Community College, offered his institution's Rock Creek Campus (in Washington County) for the purpose. Thus PCC became both the location and administrative head of what came to be called the Paul D. Lintner Center for Advanced Education, named for the late president of Electro Scientific Industries. It was based on the politically expedient and educationally necessary principle of a federation. In the fall of 1984 twenty courses were offered, and by 1986 eleven colleges were affiliated with the center. Financing was originally to come from PCC, but by 1 July 1986 it was to have broadened to include the member schools, foundations, the business community, and state government (either from the system of higher education or directly from legislative appropriations). By 1986 defenders were arguing that the combination of the Oregon Center for Advanced Technology Education and the Lintner Center was roughly equivalent to a technical university.[42]

The second feature of the business-education compact was the partnership between local businesses (again not all in high technology) and Washington County elementary and secondary schools, whose aim was to publicize technology in the community, to get students interested in the field, and to provide facilities for high-technology firms. In January 1986 the Beaverton School District appropriated forty-four thousand dollars for the business-education partnership program.

The third aspect of the compact was the establishment of internships for teachers in local firms and a pool of businesspersons and other professionals whom local schools could use as supplementary teachers in other ways. In addition, a "student career passport project" was designed in which members of the business community would identify the skills students needed to get their first jobs and the elements in the curriculum that contributed to the acquisition of those skills.[43]

OTHER JOINT EFFORTS

Besides these broadly based programs, the high-technology industry has produced a range of educational activities—some large, some small; some temporary, some lasting. Electro Scientific Industries gave Texas A&M University two ESI Model 80 laser-processing systems. Faculty and students at the university, in return, studied and developed new laser-processing applications in microelectronics circuit fabrication. Both corporation and university gained from the interchange. In time ESI opened ESI/SW, an engineering center in College Station, Texas, the home of Texas A&M. The center's first task was to conduct research in semiconductor technology and new product research. In March 1980 Floating Point Systems gave an AP-120B computer to the Department of Electrical and Computer Engineering at Oregon State University, to be used for both teaching and research. In the fall of 1983 Oregon State University and the University of Oregon invited forty-five corporate executives from around the United States to attend a meeting to apprise them of the research facilities at the two universities. The following April they launched an Advanced Science and Technology Institute to give colloquia and university internships for corporate researchers. Included were liaison arrangements, based on an annual membership fee, designed to provide corporations with research briefings, prepublication access to reports, and contact with faculty members. In 1983 a privately funded Saturday Academy was founded to provide sixth- through twelfth-grade students special classes in science, mathematics, and computer science, with a few humanities offerings. High-technology institutions that have contributed funds, faculty, and equipment to the academy include Tektronix, Floating Point Systems, Hewlett-Packard, Intel, ESI, Mentor Graphics, and the Oregon Graduate Center. By 1986 approximately 950 students had enrolled in the program.[44]

By 1986 the various universities had embarked on a range of other programs. The Oregon Center for Advanced Technology Education included the Oregon Executive MBA Program—a degree offered by Oregon State University, the University of Oregon, and Portland State University—which focused on strategic management approaches to product innovation and business development. Portland State University also began a doctoral program in electrical and computer engineering. Another development was the 2 + 2 + 2 program, in which a student took a two-year associate degree at Portland Community College, a bachelor of science degree at the University of Portland, and a master's degree at the Oregon Graduate Center. In 1982 high-technology firms, including Tektronix, Intel, Hewlett-Packard, ESI, PM Industries, and Pacific Hybrid Microelectronics, helped plan a curriculum and contributed equipment to a two-year associate-degree program at PCC in microelectronics process technology. The purpose was to furnish graduates for local electronics firms using thick- and thin-film hybrid circuits.[45]

High technology was also successful in another educational realm. When President Joseph C. Blumel resigned in early 1986 after twelve years as president of Portland State University, the chancellor appointed a search committee of faculty, students, administrators, and public members to recommend his successor. The chairman of the committee was Gene Chao, president of Metheus Corporation.

The history of the relationship between high technology and education demonstrates the power of the industry. After a long delay, it began to exercise a great deal of influence in the eighties. The results of that influence, for the long term, will depend on the success of the industry and on the public's willingness to support its educational aims.

8

Politics

WASHINGTON COUNTY'S HIGH-TECHNOLOGY INDUSTRY does not operate in isolation in the political realm any more than in the economic one. It is affected by politics at the national, state, and local levels. Depending on the specific issue and the political climate at any given time, the industry has been more or less successful in its quest for political advantages. In general, as might be anticipated, it has been more successful at the local and state levels than in Washington, D.C.

In the game of national politics, the Washington County industry is a vigorous player. In most cases, Washington County firms have been represented in their lobbying activities in Washington, D.C. by trade associations, especially the forty-year-old American Electronics Association, which includes almost three thousand companies in its membership. Frequently, corporate officers of Washington County firms journey to Washington, D.C. and they maintain close and frequent ties with the Oregon congressional delegation, other members of Congress, and administrative officials.[1] Issues that concern the Washington County firms are diverse and several, but tax policy is certainly one of the most important.

The first great gain in federal tax policy for high technology was the law of 1978, enacted against the background of a fluctuating capital-gains tax. Before 1969 the tax was 25 percent. In that year there were almost seven hundred public underwritings of small new companies and $170 million invested in new companies. But that was the year Congress raised the capital-gains tax to 48 percent. In 1975 there were only four new small companies and only $10 million invested. To restart the flow of venture capital, some high-technology officials in California formed a committee chaired by their colleague Edward Zschau, who was later a member of Congress. The committee concluded that the difference between regular and capital-gains tax rates had all but disappeared in the seventies. The committee's reasoning and lobbying were convincing and resulted in the law of 1978, cutting the tax to 28 percent.[2] By 1980 the tax was lowered to 20 percent.[3] In the eyes of high-technology corporate officers and venture capitalists, the move produced spectacularly successful results. As the authors of a book on the subject noted, "By the end of 1982, the amount of new private capital committed to venture capital firms had risen to over $1.4 billion, eight times the amount of twelve years before."[4]

The congressional session of 1981 produced a tax law that contained several benefits for high technology: a 25 percent tax credit for present R & D expenditures that exceeded the annual research-and-development outlay during the previous three years (this was in addition to the existing tax deduction for research and development); a break for employees receiving incentive stock options; an inducement to donate equipment to universities that would give the firms a deduction of the cost of the equipment plus half the difference between its cost and fair market value; and a reduction in the capital-gains tax rate to 20 percent. The following year some Washington County firms were the beneficiaries of the Small Business Innovation Development Act, designed to increase the flow of federal and R & D funds to companies with fewer than five hundred employees. The measure established a three-phase contract schedule for a recipient corporation. Phase One gave up to fifty thousand dollars for a company to determine the technical and commercial feasibility of a product planned in response to solicitations issued annually by eleven participating government agencies. Companies surviving the first phase of competition would be given Phase Two contracts of up to five hundred thousand. Before applying for these awards, candidates were urged to seek assurances of private funding for commercialization of the innovation following the project's successful completion. Phase Three involved private commercial development of the product without government funds. Although the law passed both houses of Congress by an overwhelming majority, it was opposed vigorously by two interests. Government officials declared that small business was not as successful in research as large corporations, and universities feared a reduction in their share of R & D funding.[5]

The Tax Reform Act of 1984 had little overall effect on the high-technology industry. On the positive side, the law contained individual and corporate tax provisions to aid electronics companies. But the industry initially lost heavily when the 1981 R & D tax credit was allowed to expire at the end of 1985. The industry had lobbied hard for either an extension or a permanent tax credit for this purpose. Its representatives had argued that the credit should be permanent because high technology firms needed to plan far ahead, that their technology was expensive, and that they might be forced to move their research facilities to foreign countries. Opponents, including many members of Congress, argued that the national interest required abolition of tax provisions that were revenue losers.[6] The credit was revived in later legislation.

Previous struggles over taxes affecting high technology were nothing compared with those that took place in 1985 and 1986, when "tax simplification" became a major national political issue. First raised by Democrats, it became a popular bipartisan measure after adoption by President Ronald Reagan. The idea was to reduce tax "loopholes" and the number of tax brackets. Corporation and individual income-tax rates would also be cut. What concerned high-technology firms was that the lower capital-gains tax rate would be eliminated. The argument made by the lobbyists and corporate executives who rushed to Washington was that although the capital-gains tax seemed to be a "loophole," it increased revenue because it increased economic activity at home and abroad and created new jobs.[7]

In a public statement, Thomas Bruggere, the president of Mentor Graphics, argued that the capital-gains differential had been of enormous benefit to his firm. He said that "with the help of $10 million of venture funding, Mentor Graphics grew rapidly in its first three years, and today employs nearly 450 people in Oregon and has an Oregon payroll of more than $10 million." Opponents struck back by asserting that eliminating the capital-gains differential would not reduce venture-capital investments because corporations and large investors would continue to make them; that the value of company stock would rise when corporate taxes were reduced; and that the capital-gains advantage would be retained for corporations. They also maintained that there was an overwhelming public demand for tax simplification, which few politicians or industries dared resist.[8]

High-technology firms were also concerned about several other measures in the new tax bill designed to close loopholes. The House version would eliminate the investment-tax credit for plant and equipment modernization. The high-technology lobby was disturbed by this proposal, arguing that the average life process of one of its products—from invention to design—was short, because of automation, worldwide competition, and a public fascinated with new products.

The fate of the R & D tax credit due to expire at the end of 1985 was also in doubt in the early days of the bill's history because a renewal passed the House but not the Senate.[9] The industry declared that research and development was its lifeblood and argued that it was unfair to retain tax deductions for real estate, for example, and other consumption investments, while removing those that would make the nation more competitive and productive. High-technology officials from Washington County maintained that their firms invested $300 million annually in research and development.

Local high-technology firms were also concerned that the new tax bill contained a minimum-tax provision that required every corporation to pay some tax and that companies (even those, like start-up companies, that were not yet profitable) capitalize R & D expenditures. A somewhat embarrassing footnote was the disclosure that Tektronix, the largest high technology firm in the state, had paid no federal income taxes in 1984 and 1985. Although the company's tax refunds were completely lawful, they drew adverse publicity and forced a company representative to explain "that the company's tax liability was erased because of a combination of the company's declining profits, its continuing investment in new plants and equipment, and an increase in research and development expenditures."[10]

INTERNATIONAL TRADE ISSUES

One of the most controversial issues of modern American politics is that of international trade. Because the United States is a part of the world economy, the balance of payments, deficits, and foreign competition (especially from Japan) is of great significance. Matters of exports and imports are of deep concern to the high-technology industry, which has always depended heavily on foreign markets. The question of export controls is therefore crucial to it. The departments of commerce, energy, and defense control exports to eastern-bloc countries or to nations that might in turn transfer them to the Soviet Union or its allies. Because many of their products can be used for military as well as industrial purposes, high-technology firms are particularly vulnerable to such restrictions. The Export Administration Act of 1979 was especially rigid in the matter of export licensing. As the act was about to expire at the end of 1983, high-technology interests struggled to pass a more liberal measure, a general-license category authorizing export of controlled products to noncommunist nations. The Defense Department, on the other hand, wished to review certain license applications for exports of militarily critical goods to these countries.[11]

Related to this were the distribution-license regulations of the Department of Commerce, which allowed companies with extensive international distribution to obtain a license covering multiple sales to certain overseas distributors or subsidiaries. During the era of Soviet-American detente in the seventies, these regulations were loosely enforced; but the department proposed amending them in January 1984 to stanch the flow of strategic U.S. technologies to the eastern bloc.

One proposal was to require that foreign customers of American firms agree in writing not to resell or reexport products without U.S. government approval. A coalition of politicians and industrialists from high-technology states succeeded in watering down the new regulations substantially so that they emphasized self-policing by license holders and their foreign customers.[12]

The government's concern for protecting American military high technology from illegal exports to the Soviet Bloc also manifested itself in 1981, when the Reagan administration began its Operation Exodus. Under that program, which was aimed at commercial exports in danger of diversion to military uses, the Department of Commerce issued export licenses while the Customs Service did the policing. Bureaucratic confusions between the two agencies and inconsistent enforcement of them (some U.S. departure points required licenses for certain products while others did not) raised the ire of executives from Tektronix, ESI, and other companies in Oregon.[13]

A related issue was the attempt to stimulate exports by small companies, including high-technology firms, by passing the Export Trading Company Act of 1982. Underlying the statute was the fear that the United States was losing foreign markets, in part because of having too few export firms. In 1982, 80 percent of American exports came from 1 percent of the exporters. The Export Trading Company Act was designed to make it easier for small companies to sell overseas by relaxing banking and antitrust regulations so that they could cooperate to form jointly sponsored export trading companies.[14]

Talk of cooperative ventures among high-technology firms showed up in other ways in the halls of Congress. Concern about competition from Japanese imports led to a national soul searching in the eighties. Businesspersons, politicians, unionists, and commentators speculated about the reasons for Japanese success in penetrating American markets and wondered if Americans might learn something from the Japanese system. One thing that struck Americans favorably was the pooling of R & D efforts by Japanese firms, which enabled them to accomplish faster and less costly research than was possible under individual corporate endeavors. But this aspect of Japanese business could not simply be transplanted to the United States. Those in business feared government control, while politicians worried about possible violations of antitrust laws. Attempts to steer through these shoals began in 1980, when the Carter administration issued an "Antitrust Guide Concerning Research Joint Ventures"; progressed to the introduction of several bills in Congress in 1983; and culminated in the National Cooperative Research Act of 1984, which provided antitrust protection.[15]

ENVIRONMENTAL CONTROLS

After the exposure of the leaking underground chemical storage tanks in Silicon Valley in November 1981, the high-technology industry somewhat lost its reputation for cleanliness. Since that time a great deal of money has been spent by the industry in cleaning up and attempting to prevent ground, air, and water pollution, but a striking element in the picture was the lack of federal standards for the protection of groundwater. Not that the federal government was uninterested in the problem; rather, its concern was unfocused. In 1986 sixteen federal laws dealt with the problem, but groundwater was not the sole subject of any of them. The Environmental Protection Agency (through its Office of Ground-Water Protection) was expected to be responsible, but it shared regulation of the resource with at least nine other federal agencies. In 1985 and 1986 bills were proposed in Congress to set a single national standard for groundwater cleanliness and to support local and state government attempts to restore and maintain it, but no firm action was taken.[16]

Congress did act, however, to clean up existing toxic-waste dumps, by enacting a "Superfund" for site cleanups paid for by taxes on chemical, metal, and petroleum companies. When the law came up for reauthorization in 1986, the U.S. Senate passed a bill changing the method of financing Superfund—from specialized taxes to a value-added tax on all companies, regardless of their individual responsibility for disposing of hazardous wastes. Some firms in the high-technology industry fought hard against the change, claiming that the proposed tax would disproportionately affect their operations.[17]

PROPOSED U.S. INDUSTRIAL POLICIES

Politics is the art not only of obtaining what one desires but of preventing what one does not. Again, Japanese success brought to the fore an issue with which high-technology firms were forced to grapple: the development of an industrial policy for the United States. Some Americans were taken with the Japanese government's method of targeting industries that it would help finance in order to make them competitive in international markets. Prominent American economists, such as Lester Thurow of MIT, wanted the federal government to revitalize American industry by playing a large role in planning the national economy.

The issue became specific in 1984, both in Congress and in the presidential-election campaign. Many Democratic legislators and candidate Walter Mondale, representing the interests of traditional businesses of the Northeast and Middle West, endorsed a bill introduced in the House of Representatives by Congressman John J. LaFalce of New York to create an Industrial Competitiveness Act. This act would establish three agencies to restore American competitiveness overseas. The Council on Industrial Competitiveness would create industrial-promotion policies to help U.S. businesses compete with foreign firms; the Bank for Industrial Competitiveness—based on the government's rescue of the Chrysler Corporation in 1985—would become a minority partner with private-sector banks in making loans to industries facing "serious international competitive challenges"; and the Federal Industrial Mortgage Association would provide a secondary market for long-term loans to small- and medium-sized businesses. In this session, LaFalce also introduced the Advanced Technology Foundation Act, which would establish centers for applied research at universities and help finance industrywide joint R & D ventures.[18]

Although industrial policy has received a lot of attention, it has not been translated into legislation. There was little real political support for it, especially from the high-technology industry, which argued that government should play a more modest role in aiding it. As Robert Noyce of Intel put it three years earlier, before a congressional committee: "The American response to... aggressive foreign development programs should not be to emulate the policies and practices of our trading partners... but we should seek to achieve the same objective as that of foreign high-technology programs... without direct government intervention in the free-market process."[19] Most of Washington County's high-technology business community shared his views. Their success had been enjoyed under a system of indirect rather than direct benefits from the federal government. Most were Republican. They saw little to be gained by changing a system that had given them the opportunity to succeed. But as for most interest groups, consistency is not always wise where its own interests are concerned.

A case in point is the amendment tacked onto the federal appropriations bill on 19 December 1985, wherein two Oregon Congressmen, Sen. Mark Hatfield and Rep. Les AuCoin, used their power to aid two local high-technology enterprises. The amendment declared that a parallel supercomputer developed by Floating Point Systems be the standard against which all competition for a $65 million grant for supercomputers to designated universities be measured. The second part of the amendment was to grant $1 million to the Oregon Graduate Center to buy one of the new computers. (As mentioned earlier, Hatfield was one of the early supporters of the Graduate Center.) The amendment was controversial, for in selecting universities Congress bypassed the traditional peer-review process, whereby scientists evaluate grant proposals on their merits, in favor of a "pork-barrel" process of using political influence to select the recipients of federal funds. Because of the process, Cornell University rejected the $10 million it was given under the supercomputer amendment. But the Oregon Graduate Center had a less tender conscience. As one center official put it: "One uses influence where you have it. The [congressional] delegation has done their job for Oregon, and I don't think we have anything to be ashamed of."[20]

STATE POLITICS

As one of the largest employers in Oregon, the high-technology industry has maintained a great deal of interest in state politics in recent years. It was not always so, for early in its existence high technology had relatively little to gain or lose from the actions of state government. Tektronix had maintained a lobbyist in the state capitol for some years, and high technology was represented to some extent by the Associated Oregon Industries, but in the eighties the industry's involvement in Salem became substantial and sustained. This presence arose from several developments, the most important of which was indirect: the worst depression in Oregon's modern history, caused by the decline of the housing market, agricultural overproduction, and foreign competition, all coupled with a national recession that cut into tourism. With Oregon's three major industries—forest products, agriculture, and tourism—all in decline, a desperate search for economic alternatives began. In the eyes of many, high technology became the savior of the state (one legislator called it a "golden boy"), although many of its corporate officers were more modest in statements about what the industry could do to salvage the state's future. Nevertheless, the new respect for high technology translated into an impressive series of victories for the industry in Salem.[21]

In 1982 Oregon became the first state in which the national American Electronics Association authorized its local council to establish a government-affairs program. This action was taken mainly at the request of the council's president, Thomas Bruggere. The Oregon Council of the American Electronics Association hired its first professional lobbyist—Pat McCormick, a former legislative aide in Salem and a staff member for AuCoin—shortly before the opening of the 1983 legislative session. McCormick, high-technology executives, and the Washington County legislative delegation formed a reasonably cohesive team in the session, which became one of the most contentious in the state's history because legislators had to decide where to cut state expenditures. The team decided on a small agenda: the sales tax, aid to higher education, and the state's land-use planning process.

The sales tax was, and has remained, a major goal that the industry wants from the legislature and the people. High-technology spokespersons contended that Oregon had an "unbalanced" tax system (a "two-legged stool" is a favorite metaphor) because it depended on the property tax and the income tax for revenue, but—unlike all but five other states—had no sales tax.[22] As in the case of most major political issues, the sales tax did not stand alone. Its advocates faced another effect of the state's declining income: the desire on the part of many taxpayers to have their property taxes limited or reduced. In the prosperous year of 1979, the legislature had provided property-tax relief; in the hard times that followed, the forces advocating property-tax limitation gained strength. An initiative measure to limit property taxes was defeated in 1978 and 1982.[23] By 1983 high technology's desire for state services, such as aid to higher education, was in danger of falling victim to the property-tax reducers. That year the high-technology industry decided to push a sales tax, coupled with some property-tax relief. Opponents of the tax, led by organized labor, which argued that it was regressive, were vocal and well organized when the plan came up in the legislature.

Given the depression, the historic Oregon hostility to a sales tax (one had been rejected in 1968), and the opposition of Gov. Victor Atiyeh and Senate President Ed Fadeley, it was not surprising that the tax did not pass. What was surprising was that McCormick was at least able to unite the sales-tax forces behind a single measure that, though it was defeated, dominated the session and thus gained statewide attention. As he put it, "It gave us a way to discuss Oregon's overreliance on [the] personal income tax." Or, as a legislative opponent of the sales tax observed of high technology's influence in the legislature, it was helped by the industry's "mystique. . . . it's because people look to them as the employer of the future."[24]

The American Electronics Association was also successful in the initial skirmish of what would be a long war. The Oregon Public Employees Union sponsored a bill regulating the maintenance and use of video-display terminals which would require the director of the Workers' Compensation Department to adopt rules to protect the safety and health of terminal operators and would require employers to pay for annual eye examinations for their workers. The American Electronics Association argued that the bill was expensive to industry and unnecessary, as VDTs were not dangerous to health. The bill was shunted off to the Senate Labor Committee, where it died, although the committee's chairman, Margie Hendrickson, succeeded in having an interim committee appointed to study the issue. Another measure that would arise again, "comparable worth," was also questioned by the association and opposed by other representatives of industry, such as the Portland Chamber of Commerce and Associated Oregon Industries, which bottled it up in the Labor Committee. This measure proposed that it be "declared to be the public policy of the state of Oregon that an equitable compensation relationship between traditionally female dominated and traditionally male dominated state employment be established." Moreover, the American Electronics Association was the only industrial group to seek to revise, rather than oppose outright, a workers' right-to-know bill, which would require posting notice of hazardous materials in the workplace.[25]

The momentum gained in the regular session carried over to the fall, when the governor called a special legislative session in September to address the sales-tax issue. A sales tax was passed, to be referred to the voters in the March 1984 election; but opponents, led by Fadeley, attached the condition that before it could be referred it be approved by a majority of each of the state's school districts, cities, and counties. The sales-tax campaign began, financed by a two hundred thousand dollar war chest, of which the American Electronics Association's members raised more than fifty thousand, but ended suddenly when the state Supreme Court ruled that the referral condition was unconstitutional.[26]

High technology also enjoyed success in the field of higher education in the regular 1983 session. The American Electronics Association took an unprecedented step by becoming the first group outside education to lobby for higher-education funding. The reason was not a concern for higher education as a whole, but rather the need of high technology for teaching and research facilities for science and engineering beneficial to it. Regardless of its motive, the association organized the testimony of high-technology executives before legislative committees, where they were well received. Although high technology was not the only force for an increase of 17 percent in the higher-education budget (one of the few areas to gain a substantial increase), it was a considerable one. High technology also gained another $1 million for the high-technology consortium, a sum matched by the association.[27]

In 1984, high technology could boast of a spectacular legislative success, the repeal of the unitary tax, although not all forces behind it were Oregon's high-technology firms. The unitary tax is not a tax, but a method of apportioning the state corporation income tax on multistate and multinational corporations. As one publication explained, the state "computes the total earnings for such corporations or groups of corporations on a worldwide basis. Then it determines Oregon's percentage of worldwide sales, property, and payroll. By multiplying the average of those factors times total worldwide profit, the state fixes the earnings against which Oregon levies its 7.5 percent corporate income tax."[28] Foreign corporations and multinational Oregon corporations, such as Tektronix, argued that the unitary tax was unfair in taxing profits earned elsewhere. Foreign firms did more than argue; they threatened. A Japanese business delegation visited Washington, D.C. in June 1984 and was given an audience with President Reagan. It then visited the unitary-tax states, threatening as it went that no new investment would come to these states unless the tax were repealed. By the fall of 1984, the secretary of the treasury was also threatening the states. He declared that unless the states repealed the unitary tax by July 1985, the administration would "recommend federal legislation that would limit the reach of state corporate taxes to a company's earnings within the U.S."[29] With Oregon's economy in depression, the governor hastily ordered a Department of Revenue study of the effects of the unitary tax on encouraging business to come to Oregon. The mayor of Portland called for its abolition, the State Senate Trade and Economic Development Committee scheduled hearings on the tax, and Congressman Ron Wyden proposed a bill that would forbid states to tax foreign earnings of corporations that conduct at least 80 percent of their business outside the United States. Local high-technology people were enthusiastically in favor of repeal.

The threats and studies worked. The governor, who had received a good deal of information on the effects of the tax from the electronics industry, especially from a Tektronix tax expert, became convinced of the wisdom of repeal.[30]

On 20 June 1984 Atiyeh made a speech to the Portland World Affairs Council to declare his intention to ask for the repeal of the unitary tax. "The Japanese, our neighbors on the Pacific Rim and Oregon's leading trading partners," he warned, "say 170 of their companies are looking for sites for new plants. But they have told us bluntly their corporations will not even consider a state that employs the unitary method.... That is a grim warning to Oregon, which remains one of the minority of states using the unitary method." At the time the governor did not call for a special session of the legislature to repeal the tax.[31]

High-technology officials were among the first to endorse the governor's proposal. A Tektronix official said, "We're doing the right thing and being rewarded for it in new investments." Senator Jim Gardner, chairman of the Senate Trade and Economic Development Committee of the 1983 legislature, and Grattan Kerans, Speaker of the House of Representatives, were among the important political leaders who endorsed the plan. The Japanese, albeit discreetly and indirectly, entered the debate. Visiting businesspersons told Oregon business and political leaders that "if Oregon expected major plants to be located in this state from Japan, it was going to be absolutely necessary that we do something about the effects of the unitary tax formula."[32]

Not all favored repeal. Critics abounded, led by Irv Fletcher, president of the state AFL-CIO, who argued that the measure would cost the state assured income in exchange for the mere promise that "a couple of companies" would come to the state. Many others feared that the loss of state revenues would mean the reduction of social services. Others argued that multinational corporations could hide the Oregon portion of their income through "creative accounting."[33]

In early July, Atiyeh came out in favor of a special session to repeal the unitary tax as well as recently passed income-tax increases on nonresidents (principally Washingtonians) who worked in the state. Editorial writers and other political leaders feared a special session. They were afraid that the legislators, once summoned to Salem, would turn the session into a wide-open affair on a host of subjects that could drag on for weeks. Even some who favored repealing one or both taxes said the job could wait until the January regular session of the legislature, especially since the tax changes would probably not take effect until the end of June 1985. In the end, however, the governor was able to get the consent of legislative leaders to hold a special session and to limit its agenda to the two tax matters.[34]

The governor and his political and business allies won an outright victory. Well before the special session met they reached a consensus with legislators to pass the two tax measures. It was thus something of an anticlimax when the legislators met in Salem on 30 July. The House voted 53-6 to repeal the unitary method; the Senate 27-3. Several who voted for repeal felt pressured to do so; the chairperson of the Senate Revenue Committee stated: "I join the group that holds their nose and kicks the desk and votes for this bill." In only thirteen hours the legislators had come, acted, and departed. Oregon became the first of the twelve unitary-tax states to vote for repeal.[35]

One veteran legislator marveled that a complex tax with a large revenue impact would be repealed so quickly, and credited it to the high-technology lobbyists:

> The high-tech industry was very skillful in presenting a very accurate case. . . . They were able to lay it out and get the action taken in an unbelievably short period of time, because it was tied into the fact that if we changed this method of taxation, and it made Oregon more attractive for high-tech industry, people were really willing to buy that even without going through the really thorough examination in the long period of time we usually spend on tax issues like that. So that was really amazing. I think that would really signify their tremendous influence and their tremendous power.[36]

The vote attracted much attention in Oregon, around the country (it was covered in an article in the *Wall Street Journal*), and in Japan, where at least three financial newspapers and two major banks hailed the repeal. One Oregon editor, on the losing side, dashed off a poem on hearing the news:

> Give me your rich, your strong
> Your corporate giants yearning to be free,
> The wealthy companies of your teeming shore.
> Send these, the high-tech homeless to me.
> I lift my lamp beside the golden door.

Time would decide which view was justified.[37]

High technology's momentum gained in the 1984 special session continued into the fall elections of that year and into the regular session of 1985. The last two successes were related. After a dispute with the national American Electronics Association, which opposed the idea, the Oregon Council finally received permission to establish a political action committee (PAC). It was decided to spend the PAC's money on legislative elections almost exclusively. The local American Electronics Association committee interviewed the candidates in thirty contested elections and eventually gave a total of almost fifteen thousand dollars to more than forty candidates. The largest sum dispensed to individual candidates was one thousand dollars. Two of the beneficiaries of the maximum donation were Rep. Vera Katz and Sen. John Kitzhaber, who had been friendly to high technology in the 1983 session. They were worth investing in, for in the 1985 session Katz became Speaker of the House and Kitzhaber president of the Senate. Overall, 85 percent of the candidates supported by the PAC were victorious. After the votes were counted, the PAC judiciously dispensed its remaining funds to support some winning candidates it had not supported earlier and to help pay the campaign deficits of winners who were in important legislative positions. It should be noted that American Electronics Association funds were not necessary to gain legislative support from most legislators, as many—by conviction, employment (IBM, Pacific Northwest Bell), or geographic district (Washington County)—were already safely in the high-technology fold.[38]

When the legislators met at Salem in January 1985, the industry was on the verge of almost total success: during the session it lost on no major issue. Its greatest triumph was in the passage of the sales tax to go before the people in November. In this campaign the first victory was in helping persuade Governor Atiyeh to endorse the tax. Like most veteran public servants, the governor had invested much time and effort in developing the existing tax structure. With Katz, Kitzhaber, the American Electronics Association, and others he designed a plan that would appeal to industry and to the proponents of property-tax reduction. The sales-tax agreement that emerged from the session was complex, including a constitutional amendment and statutory changes.

When the campaign was taken to the people, the American Electronics Association and the high-technology industry were in the forefront. The governor, who chaired the campaign, included Pat McCormick as one of the five members of its executive committee. Norman Winningstad of Floating Point was the most visible high-technology advocate of the sales tax, flying about the state in his helicopter to address meetings, engaging in debates with opponents, and giving interviews. The American Electronics Association and its members contributed substantially to the campaign, in money as well as time. But to no avail. The tax was overwhelmingly defeated by a margin of 664,365 to 189,733. In the aftermath, political analysts explained the debacle in a variety of ways: Oregonians' traditional hostility to the sales tax, as well as the perception that it would raise the voters' tax burden rather than redistributing it, that the "package" helped the rich and poor but not the middle class, that the campaign was incompetent and insulting (it included a comic book), and that the tax by its nature was regressive.[39]

In the 1985 legislative session, the American Electronics Association won a second victory in the area of state funding of higher education. The Oregon State System of Higher Education had convinced the governor to recommend a substantial increase in its funding: a $40 million budget for salaries, lottery funding for higher-education buildings, and special funds for "centers of excellence" at the three state universities. Originally, legislative leaders were not enthusiastic about the proposals. Some wanted to cut the salary request by $10 million or $20 million and reduce building funds by two-thirds. The American Electronics Association, however, worked with the Oregon State System of Higher Education to form teams of high technology officers and university officials to call on legislators. These visits, in addition to support from faculty organizations, engineering and legislative alumni, and ordinary citizens, helped contribute to the passage of the entire state system requests, the first substantial gains for higher education in several sessions.[40]

The American Electronics Association was also an important contributor to the development of the Oregon Center for Advanced Technology Education. By 1985 the Consortium for High Technology Education had run its course and the governor's recommended higher-education budget had reduced the state system's proposals for engineering programs at Portland State. The American Electronics Association proposed the idea of the Oregon Center for Advanced Technology Education, which immediately received the support of Governor Atiyeh and Chancellor William E. "Bud" Davis. There was some initial legislative reluctance, especially from speaker Katz, who feared that the center would weaken the consortium, but the staff of the association and the chancellor convinced her that it was dead in any case. Thereafter, the legislature became more enthusiastic, especially when Atiyeh promised to urge the Ways and Means Committee to shift the $1 million earmarked for the consortium and the $300,000 left from the Council for Advanced Science and Engineering Education/Research for Industry to pay for the center. This lack of an additional expenditure convinced the committee and, ultimately, the legislature. The American Electronics Association also arranged a dinner meeting with several high-technology executives and two members of the Ways and Means Committee's Subcommittee on Lottery Expenditures, which paved the way for the center to be furnished with some lottery funds.[41]

Another measure, resurrected from the 1983 legislative session, was that of video-display terminal regulation. High technology dug in its heels on this bill. Although the Senate Interim Committee Task Force reported a much milder bill than had been presented in 1983, the American Electronics Association remained adamantly opposed. Atiyeh, by then the association's faithful ally, asked it to help draft an executive order to deal with workplace design and office technology in order to head off the bill. The association cooperated, but the bill passed both houses because legislators knew that the governor—who, unlike them, could not run for reelection and could thus risk alienating women's and labor groups—would veto it as he had promised.[42]

Also back from 1983 was the issue of environmental protection. Organized labor, environmental groups, public-interest lobbies (such as the Oregon Student Public Interest Research Group), and many ordinary citizens were concerned about radiation, underground water pollution, and transportation of hazardous waste to disposal sites. Among other measures that they wanted was a community right-to-know bill that would require corporations to inform the public of the whereabouts of hazardous materials. They threatened an initiative measure to take the issue to the people if the legislature did not pass such a bill. High-technology representatives opposed this bill, arguing that it would be expensive, that it might reveal trade secrets, and that information in the wrong hands could lead to sabotage. But it did not fight adamantly, as it had in the case of video-display terminals and comparable-worth bills. Public support for right to know was too strong for that treatment, and the bill passed.

The American Electronics Association was also concerned with a community right-to-know bill introduced by the Department of Environmental Quality (DEQ), which would have made the siting of hazardous materials public knowledge. It met with business groups, such as the Associated Oregon Industries and chemical industries, and convinced them to take a positive rather than a negative stance. Rather than digging in its heels, it favored introducing a bill that required corporations to inform fire and police departments of the location of hazardous materials, so that the agencies could protect their communities in case of an emergency. Careful cultivation of legislators paid off when the bill was sent to the House Environment and Energy Committee, chaired by Rep. Mike McCracken, who had received a major contribution from the American Electronics Association PAC funds and whose political campaign had been shaped by McCormick. By that time the association had made sure that the various emergency-response agencies were apprised of the bill's contents and were in support of its approach. McCracken asked Rep. Mary Burrows to head a group of concerned interests to work out a bill that would be certain of passage. The result was the Community Right to Know and Protection Act, which required the state fire marshal to conduct annual surveys of businesses to detect hazardous wastes and to inform the public of the results. Police and fire departments and hospitals could use the information in an emergency. The bill passed both houses almost unanimously and was signed by the governor. On a related matter, the association—without arousing public opposition—was able to kill a section of the bill that would have required every company to indicate where it had disposed of its hazardous wastes since 1952.[43]

The American Electronics Association won several other battles in the environmental realm. The DEQ had proposed a bill giving it the authority to charge owners of leaking storage tanks the costs of investigating and cleaning up the leak. Congress had passed a law in 1984 giving the federal Department of Environmental Quality authority to set standards for underground waste disposal, but the department had not done so. The association was able to amend the bill to have the state set standards for new tanks without waiting for the federal rules and to prevent local governments from developing their own standards. The bill passed. In alliance with title companies, the association won another victory when the DEQ proposed a bill requiring that notices be placed in deed records of properties containing potentially hazardous environmental conditions. The American Electronics Association feared that such notices would remain forever (as deed records are permanent), even if a site were cleaned up. It was able to water down the bill to provide a land-use overlay on such sites, limiting their potential use and ensuring that adjacent landowners would have sufficient warning. The association was also able to dilute a DEQ bill that would have taxed feedstock to pay the costs of training and equipping emergency-spill response teams around the state and that would have used the money to match federal Superfund dollars in cleaning up eligible sites in Oregon.[44]

From the first year it formally made its presence known in Salem, the American Electronics Association had enjoyed considerable legislative success. Its major goals in the areas of tax, educational, and environmental measures had all been achieved. The causes for these successes were several. First, the public, the legislature, and the governor hoped that high technology would lead the state out of its economic difficulties. Second, the association built alliances with other interest groups to promote its own program. Furthermore, it tried to move in a positive manner, to be flexible, to adjust, to amend, to compromise, rather than to oppose outright. Finally, it was more than a lobbyist. Executives of its member firms (eighty-three in number representing 85 percent of the state's electronics industry) spent much time and labor in gubernatorial, legislative, and state agency offices to advance their causes.[45]

ECONOMIC-DEVELOPMENT GAINS

Aside from its legislative achievements high technology gained from state government one of the major political and economic issues of the 1980s, economic development. The concern, at least on a limited scale, went back many years. In 1959, for example, Governor Hatfield had led a government and business delegation to the Western Electronic Show and Convention at San Francisco in an attempt to lure new industry to Oregon. It was in Hatfield's administration that a state planning-and-development department was created. In the early 1960s editorials hailed the growth in electronics employment. Representative Edith Green, Robert Short of Portland General Electric Company, and President James Jensen of Oregon State University went to Washington, D.C., to try to persuade NASA to situate a $55 million electronics center in the Tualatin Valley. Sporadic efforts at the state level continued into the 1960s, when Tom McCall became governor. McCall articulated the Oregonians' philosophy of that decade when he urged visitors to "come but not stay." Although the remark was never intended to apply to businesses, it received a lot of national publicity and may have contributed to a perception that Oregon was "anti-business." Actually, McCall was successful in proposing that the legislature establish an Economic Development Department in 1973.[46]

By the depression of the late seventies, a push for state programs to aid economic development resulted in the special session of 1982, when the legislature created the Oregon Consortium for High Technology. It also appropriated its first sum of money ($250,000) specifically for industrial recruitment, to be spent on marketing and promotion. In the 1983 session the legislature appropriated $13 million for the state Economic Development Department. It appropriated $2.6 million for high-technology education at the state's three major universities and another $1 million for the Educational Coordinating Commission, which would be matched by the Oregon Council of the American Electronics Association. It aided construction of new industries by eliminating the delay in getting the necessary local government permits—hitherto one of the stubbornest obstacles. Under the new law, local governments were required to act on completed development applications within one hundred days, and appeals made against these decisions were to be made in less than seventy-seven days.[47] In 1983 the legislature created a Commission on Futures Research to "study long-term economic development with focus on developing an integrated state policy,... to produce a vigorous economy in the first decade of the twenty-first century." The commission, under the leadership of Douglas Strain of ESI, reported in 1985.[48]

In 1984, under a new director, John Anderson, the Economic Development Department focused on bringing outside businesses (especially high technology) to Oregon and expanding international trade. These goals were not necessarily compatible, and they divided the state's business community. Anderson had to straddle the issue, saying: "My own view is that we look toward attracting high technology investment. . . . That we should do it with a goal of using and selling and promoting Oregon as an export platform."[49]

Whatever was principally responsible, 1984 also brought concrete returns from the governor's missions: the repeal of the unitary tax, legislative appropriations for higher education, and (with the efforts of the Economic Development Department) one new corporation—NEC.

Few were complacent about these gains, however. Every state was trying to draw new high-technology firms, and almost every state offered one or more of the following inducements: direct investment of state funds as venture capital in new companies; tax credits for venture capital; inexpensive office and laboratory space for new companies; grants of state money for research and development; state funds for "technology centers" near universities; increased funds for higher and secondary education. Oregon's difficult challenge was that the state was well behind others—such as California, Massachusetts, Texas, and North Carolina—that had developed a high-technology "critical mass," the infrastructure and camaraderie that would continue to develop capital and bring new engineers to the state. One outside commentator argued: "Infrastructure is so important in high tech that success will breed success. The rich will get richer."[50]

To make a virtue of its late start and limited state resources, Oregon's Economic Development Department had to rely on more traditional means of attracting high technology. It had to market the state without offering much by way of direct state aid to new industries. Thus it touted cheap land and housing, clean water, an educated work force, natural beauty, and the intangibles that make up an attractive way of life. Department officials had to argue (unconvincingly) that direct state aid was unnecessary, that there was plenty of private venture capital, that the state educational institutions fostered cooperation between universities and industry, and that private real-estate developers had adequate facilities for new corporations.[51]

In 1985, the legislature directed the Economic Development Department to undertake an industrial marketing program to draw businesses to the state. Among other tasks, the program was to identify "the industries, industry segments, geographic regions or other groups and factors to be targeted by the marketing program." Further, the department's director was to submit a biennial report to the legislature on the program.[52] In the same session, the legislature passed a law permitting counties to develop and implement a program of economic development that would "clearly state community economic goals and provide a detailed program to carry out these goals."[53]

LOCAL GOVERNMENT

The influence of high technology may also be seen in local government. It has evolved from a modest concern—almost a lack of concern—to a major role in Washington County politics. When Tektronix moved to Washington County in 1951, the county's population was 61,269, while Hillsboro's was 5,142 and Beaverton's 5,937. In those placid days, Tektronix discouraged its members from political activity. One county politician knew a Tek employee who wanted to run for the state legislature but was told, "If you want to run, you resign."[54] By the late fifties and early sixties, however, Tektronix had become more active in political life. Chuck Frost, a company employee, became president of the Beaverton Chamber of Commerce in the late 1950s.

One issue that concerned the company was relations with the City of Beaverton. When Tektronix moved to its new plant on Jenkins Road in 1957, there was considerable talk that Beaverton, Cedar Hills, and the new property might merge into one city. Tektronix needed sewer service and wanted Beaverton to supply it. The city would do so, but only at the price of annexation. The company was unenthusiastic about paying Beaverton city taxes, so it decided to build its own sewage plant and to reject annexation, a position it has held to subsequently whenever talk of expanding Beaverton occurs.[55] Another important issue for Tektronix was access to its property.

In the late fifties Tek got Washington County to extend Jenkins Road from its property eastward to Cedar Hills Boulevard and was instrumental in the county's extension of Ecole Road from Jenkins Road north to Walker Road. In the early sixties ESI had trouble accomplishing zoning changes in the Sunset area so that it could establish its Washington County plant.[56]

By the early seventies Tek and other firms were assuming visible political roles. One school board member of that period remembers: "It impressed me that whenever the school district was going to do anything that involved citizen participation, we always looked to people from Tektronix to chair committees... they participated very strongly in the educational process out here, as very good corporate citizens.... In fact, I don't think we would have passed budget elections without them." One secret of high tech's successful defense of its interests at the county level is its consistent presence. "They're always there, they're always serving, and they're always involved," recalls one local politician.[57]

The necessity of being involved in local government is understandable given the nature of the industry. It requires a sophisticated infrastructure, especially when new companies are beginning or old ones expanding. Land, sewers, and transportation all must be in place. Furthermore, high technology requires a solid educational system, both to produce an educated work force and to appeal to prospective employees who want good schools for their children. While these needs were obvious to high-technology executives, they were not so apparent to many citizens, especially those in the western, rural regions of the county. Nor was it always clear to officials of county government or of state agencies whose work impinged on county matters.

From the vantage of high technology, the difficulty in obtaining its objectives stemmed not only from citizens and opposing interests motivated by fear of overpopulation or nostalgia, but from the intricate structure of modern state and local government. In the seventies and eighties, high technology in Washington County had to deal with the local county government, the Metropolitan Service District, the Unified Sewerage Agency, the Beaverton School District, the Tri-County Metropolitan Transportation District of Oregon (Tri-Met), and the state Land Conservation and Development Commission

(LCDC). The political structure was also affected by the changing times: in the sixties and seventies Oregon adopted a range of measures to protect the natural environment (of which the formation of the LCDC was the most important) and its citizens took a limited-growth approach to industrial development. By the early eighties, with the state economy in depression, citizens were eager for industrial growth, although they still hoped to accomplish it without spoiling their natural environment and losing their social amenities. Local politics still reflected the historic reluctance of Oregonians, despite their progressive reputation, to spend many tax dollars for community purposes.

The most visible agency of local government was the Washington County Board of Commissioners. Traditionally, high-technology people had often been frustrated by the slowness and ineffectiveness of county government. However, by the 1980s, they found board members interested in developing a modern infrastructure, although the board was unable to persuade county voters to establish a modern tax base. (Washington County's tax base dated back to 1916.) Under Oregon law the tax base can be raised only 6 percent per year without a vote of the people. Between 1950 and 1986 the population of the county had quadrupled, yet in 1986 the tax base remained at $4 million. The difference was made up by special levies of short duration, time-consuming to pass and specialized in purpose. It is surprising—indeed remarkable—that the board or other agencies of county government were able to achieve any modernization measures at all.

If the industry was thwarted by the taxpayers' reluctance to fund services, it was pleased by the board's willingness to direct the county planning department to clear the way for zone changes, building permits, and other governmental requirements for industrial development. It is not too much to say that high technology, so far as county government is concerned, has been given a green light to develop the county for its purposes.

The cost of this development can only be guessed at. One company president says that although there is a good deal of talk among high-technology people about not wishing to repeat the Silicon Valley in Washington County, few—if any—of the industry's leaders are doing anything about it.[58]

The industry also has had reason to be pleased that other problems have been solved for it, among them the question of sewage disposal. By the late sixties the Tualatin River watershed was mired in sewage. Twenty-six treatment plants served the area, but the river and its tributaries had become receptacles for partially treated sewage and a menace to health. So dangerous were conditions on 26 September 1969 that the state Department of Environmental Quality banned new connections to sewers in the entire county except for schools, hospitals, and health hazards. The prohibition raised howls of protest from developers, real estate interests, and citizens of a county that was growing by six thousand persons a year. A citizens' committee formed to address the problem made recommendations that led to a state law authorizing a regional sewerage agency. County voters adopted the district (the Unified Sewerage Agency) in February 1970 and two months later approved a $36 million bond issue to finance the first ten-year capital-improvement plan, the largest bond issue ever attempted by a local government in the state. With its financial base established, by July 1986 the agency had constructed four major treatment plants and built or supervised 550 miles of sewer line. It was so successful that it was awarded the Municipal Water Protection Award in 1978 by the Pacific Northwest Pollution Control Association. By 1986 sewage facilities were deemed adequate for residential and industrial development in the county.[59]

On the related issue of solid-waste disposal, high technology showed its political power. In 1984 the Metropolitan Service District began the process of selecting a waste transfer and recycling site. Through legislation, the City of Beaverton prevented the site from being within the city limits. Metro then planned to put it in the heart of the Sunset Corridor. Led by Planar Systems, which claimed that heavy trucks would injure its delicate manufacturing process, and Fujitsu, which argued that the site was bad for its image, the industry appealed to Governor Atiyeh, who appointed a task force to consider another site in Reedville. The Washington County Board of Commissioners also blocked the Sunset Corridor site. Metro then had to cast about to find still another location, which, clearly, will not be near any area that high technology wishes to defend.[60]

Schools were another success story. Although voter levies did not pass as automatically in the eighties as in the sixties or seventies, Beaverton School District No. 48 was one of the better districts in the state. Upper- and middle-class people who might be scornful of many government programs backed school-bond issues, with high-technology representatives often in the vanguard. They wanted good schools for their own children and needed them to draw employees to their firms. However, by the eighties, county voters had become recalcitrant about school support. Although a tax base was passed in 1980, another was defeated in 1984 and still another in 1986. The schools survived on a series of one-year levies. In spite of voter hesitation, the Beaverton School District had sufficient accomplishments to appeal to high-technology firms and cooperated in its part of the Business-Education Compact.

If high technology was attracted to Washington County in part because of the availability of land, it recognized that it needed that land developed in an orderly manner. Land-use planning, in other words, was useful not only to the larger community but to industry. Few in Washington County's high-technology industry wanted to repeat the overcrowded sprawl of California's Silicon Valley, and land-use planning preceded all but the earliest of their enterprises. In 1954 the county's voters created a fifteen-square-mile area, the East Washington County Planning and Zoning District, governed by a five-member elected board. One of the board members was a Tektronix engineer named Henry Hayes; another early member was John Lidke, also a Tek employee, both of whom were exceptions to the company's initial reluctance for its employees to participate in public affairs.[61]

By the seventies land-use planning was a much bigger issue. In 1973 the state legislature created the Land Conservation and Development Commission (LCDC), which has as its administrative arm the Land Conservation and Development Department. Among other things, the LCDC required cities and counties to file comprehensive land-use plans with its office. Washington County government finished compliance in 1985, by which time most cities in the county were also in compliance.

Land was also the purview of the Metropolitan Service District, which had been formed in 1970. Its duties included the preparation of an urban-growth boundary to define the distinction between rural and urban lands and to identify the area in which urban development could take place over the next twenty years. In 1974, the LCDC adopted a regulation requiring all urban areas to define an urban-growth boundary. In 1979 the state legislature passed a bill introduced by Rep. Jane Cease of Multnomah County, assigning the task of establishing the Portland region urban-growth boundary to Metro. In 1986 Metro made its first major change in the boundary in Washington County to accommodate the wishes of high-technology enterprises for expansion.[62] Some high-technology industries have wanted to breach the regulations, but Metro has held firm.

Transportation was another issue vital to the industry, especially to new companies. Nowhere is the transition from an urban to a rural county more evident than in the need for a modern highway network. Between 1960 and 1986 the number of passenger cars on Washington County roads quintupled; in 1985, 62 percent of Washington County residents commuted to jobs in the county. The enlarged population required new transportation arteries, both north-south and east-west, for the entire community. High technology stressed that the state highway-tax formula discriminated against a growing region such as Washington County. Under the state formula (66 percent of state highway funds were spent on state highways, 21 percent on county, and 13 percent on city), urban residents were counted twice (as both city and unincorporated county dwellers), while residents outside the urban-growth boundary—about one-half of Washington County's population—were counted once. As a result, city dwellers got more highway dollars than did nonresidents, though the latter occupied areas that, for all intents and purposes, were identical to cities in their need for services. Above all, what the industry wanted was for county and state highway departments to have clearly defined responsibilities and adequate financing. To that end, high technology joined a coalition of local political and economic development groups (the Sunset Corridor Association) that sought to change state law so that the highway-distribution formula acknowledged the number of state highway miles in an area.[63]

Another special district with which the industry crossed swords was Tri-Met, the Portland metropolitan area's transit system. The industry fought the payroll tax on employers that was Tri-Met's lifeblood. Tektronix was a plaintiff in a 1970 suit in the Oregon Tax Court alleging that the tax was unconstitutional, but the court refused to take jurisdiction.[64] Officers of high-technology firms also objected to the planning and routing operations of Metro and Tri-Met. Especially objectionable to them was the talk of running a light-rail (electric train) system from the Tualatin Valley to downtown Portland. They contended that most people in the industry were not commuters in the traditional suburban-central-city mold, but moved from suburban home to suburban office or plant. High technology wanted Tri-Met to provide small buses for those riders.[65]

Another local governmental entity to affect Washington County's high technology enterprises was the Portland Development Commission. Founded in 1958 as an urban-renewal agency, the commission took on its economic-development functions in 1981. Not confined to the city of Portland, the agency serves the entire metropolitan area, publishing brochures on such topics as high technology, education, local government and taxes, utilities, labor and employment, real estate, transportation, demographics, and the quality of life. It has also furnished Japanese firms with information about the region and its economic possibilities, and it has probably helped to bring new firms to Washington County.[66]

In the years since the sixties and seventies, high technology's relationship with local government had taken a new direction. Most company officials now speak highly of the cooperation of local government officials in resolving their political problems. Not all that is asked is granted, but there is a cooperative political-industrial spirit at the local level, as there is at the state and national levels—a tribute to high technology's success in politics.

9

The World of Work

CRUCIAL TO THE SUCCESS OF ANY BUSINESS is its labor force. This is especially true of the electronics industry, in which, to remain competitive in the hotly contested markets for its products, employers must secure a bright, talented, motivated, stable, loyal, and—above all—productive corps of workers.

The presence of such workers in Washington County is attested, in the years from 1975 to 1982, by a growth in employment in categories labor economists term Group II and Group III industries. These are concerned with companies having the highest concentrations of "technology-oriented workers"—engineers, mathematical experts, physical scientists, science technicians, and computer specialists.[1] Group II industries are those with a ratio of R & D expenditures to net sales that is at least twice the average for all industries. Group III consists of manufacturing industries with a proportion of technology-oriented-workers equal to, or greater than, the average for all manufacturing industries, and a ratio of R & D expenditures to sales close to, or above, the average for all industries.[2]

Oregon's Group II employment, as a percentage of total nonagricultural employment, grew by 10 percent from 1975 to 1982, ranking it seventh in the nation.[3] In the same period, Group III employment rose 18 percent for an eleventh-place ranking.[4] Under the definitions used in arriving at these rankings, Oregon's high-technology industries are concentrated in the Portland area, with the majority in Washington County, where their rapid expansion has put a premium on excellent workers.

THE ATTRACTIONS FOR WORKERS

High-technology companies must root themselves in a locale favorable to their labor force if they are to succeed, because engineers and scientists—in great demand but short supply—have no reason to hire on with companies in unpleasant areas with few amenities. At the same time, the pool of hourly wage workers needs to be intelligent, motivated, and willing to work for rates lower than those earned by their unionized counterparts. Generally this has meant that companies must settle in areas with better-than-average educational facilities at all levels, a large pool of nonunionized workers, and agreeable surroundings.

Washington County has clearly addressed these requirements to the industry's satisfaction. Although there has been some difficulty in recruiting engineers from Silicon Valley, those from other parts of the country have been more willing to move to Oregon because of the state's reputation for beauty, its range of recreations, its informal style of living, its good schools, and its homogeneous population interested in middle-class values.[5] In addition, with the advent of the Oregon Center for Advanced Technology Education, the strengthening of the engineering curriculum at Portland State and Oregon State universities, and recent improvements in the Oregon Graduate Center, engineers in the greater Portland area have had more opportunities to improve their skills, receive research support, and pursue advanced degrees than previously. By the same token the county, with the highest per-capita income of any in the state, has built a respected school system at the elementary and secondary levels.[6]

Thus, by 1980, Washington County seemed poised for a great boost in its work force. It appeared that this would be accomplished through an infusion of new electronics firms. Unfortunately, the reality was quite different.

THE CORNUCOPIA MYTH

In 1986 it could not be gainsaid that high technology was a boon to Oregon's economy and labor force. But the widely held view that the industry was a cornucopia of employment was belied by several facts, most dramatically a major reduction in force announced by Tektronix. On 21 May 1986, the state's largest employer announced a cutback of two thousand positions, of which 85 percent were to come from Portland-area facilities.[7] This number was in addition to approximately twenty-five hundred high-technology jobs that had been lost the previous year throughout the industry, mostly in Washington County.[8]

Behind Tek's startling news was a trend of declining employment in Washington County's electronics industry that had begun in 1981. From 1978, when employees numbered 17,738, to 1980, when they numbered 22,318, the high-technology labor force had grown steadily.[9] In every year after 1980, however, it had declined, until in 1984 (the last year for which total Washington County figures were available) it numbered 19,159.[10] This reduction occurred primarily in the area of instruments and related products, which shrank from 17,727 in 1980 to 12,951 in 1984.[11]

Part of the reduction may be attributed to Tektronix's shift of some production from Washington County to Vancouver, Washington.[12] Nevertheless, the severe depression affecting the industry worldwide accounted for the decline in instruments employment and halted the huge growth of the county's employment in electronic equipment and supplies in 1985.[13]

This growth, credited to the establishment of semiconductor manufacturing in the county, had begun at 2,116 in 1978 and peaked at 6,208 in 1984.[14] Yet even at Intel, the county's first semiconductor firm and its second largest private employer, the payroll shrank from

4,000 to 3,250 in 1986.[15] That figure itself was deceptive, for—through rounds of new hirings followed by layoffs—it involved the loss of 1,200 workers in 1985 alone.[16]

Moreover, the outlook was not encouraging for that segment of the industry. The trend, at least among commodity-chip makers such as Intel, was to situate plants "offshore"—in Third World countries, where workers earned dramatically less than Americans could doing tedious assembly-line jobs.[17] According to state employment specialists, companies also turned to temporary agencies for workers to do "everything from assembly jobs on up to specialized engineering tasks."[18] In other words, many—perhaps most—of these jobs were not coming back as the times got better.

As it happens, the future shape and size of the high-technology work force may not depend at all on the export of jobs or on temporary workers, but on robots. By the eighties, Tektronix had already installed a computerized system to manufacture graphics terminals, representing one of the most advanced techniques available for inserting integrated circuits.[19] In the system, human hands do not touch the etched circuit boards until the entire production process is complete.[20] If, as hoped, production could be linked directly to computer-aided design and computer-aided manufacturing equipment, the linkage could put the whole process, from concept to manufacturing, into the hands of one engineer.[21] The ramifications of this development for the work force are clearly immense.

By early 1986, most laid-off electronics workers had apparently recognized these facts. While some tried to find jobs at new plants such as NEC America, Epson, and Fujitsu, many others began to filter into the trade and service sectors.[22] Those areas, and not electronics, represented the area's real job-growth categories.[23]

The cruelest aspect of the high-technology employment myth was the early belief that the industry could offer long-term, stable growth that was prey to none of the volatility associated with forest products. In reality, in the late seventies and early eighties the industry was a victim of worldwide and national economic downturns and, in 1985, of at least one internal recession induced by glutted markets.[24]

Wise heads in the press, the economic and political establishments, and the industry itself had counseled that electronics always was and always would be both cyclical in nature and only a small part of the solution to the state's economic problems.[25] The heedless, however, had accepted at face value the hoopla surrounding start-ups, plant expansions, and new plant acquisitions. The subsequent travail of the laid-off high-technology work force did not destroy the industry's mystique, but it did offer a bracing corrective to the myth of its perpetually expanding fortunes.

THE QUESTION OF UNIONIZING WORKERS

One aspect of the industry's relationship to its work force is, however, solid fact: while electronics is definitely not recessionproof, it is unionproof. The reason this is so has to do with the nature of the industry, the state, and the county.

As late as 1959, Oregon was the fifth most heavily unionized state in the nation. Although that position has since deteriorated, it still has an average of twice as many organized workers as the average for every other state. In the past, Oregon had a strong tradition of collective bargaining in a variety of industries, but the large number of organized workers could be attributed mainly to the almost total organization of the forest-products industry. Then, beginning in 1979, in the wake of the worst recession since the Great Depression, the American housing market crumbled and forest products collapsed with it. As a result, employment fell, as did the proportion of unionized workers, until Oregon began to follow the national trend where unionization was concerned. Nationally, 1986 union membership in the nongovernmental work force was at 17 percent, in contrast to a high of 35.3 percent in 1953.[26]

The rapidly accelerating disdain for unions among American workers has nowhere been so apparent as in the nation's electronics industry. While organized labor, led by the Communications Workers of America, has tried diligently to penetrate the industry, it has been strikingly unsuccessful. According to the American Electronics Association, in 81 percent of the elections held between 1977 and 1982, union representation was refused by workers.[27] Between 1972 and 1977, unions were rejected 75 percent of the time.[28]

Seen in this light, the failure of unions to organize workers in Washington County's high-technology plants seems understandable. (The Communications Workers of America have a collective-bargaining contract with only one firm in the county.) Yet the real reasons for the situation go deeper than a general trend against union membership and the effects of recession suggest. They go to the heart of the industry, to its values and its self-image, which, in turn, have conditioned the nature of labor-management relations in the American economy generally during the latter half of the twentieth century.

The entrepreneurs who built the high-technology firms of the late-postwar era—such as Robert Noyce and Gordon Moore of Intel, William Hewlett and David Packard of Hewlett-Packard, and Jack Murdock and Howard Vollum of Tektronix—were uniformly men of the lower-middle and solid-middle classes. For them, the Depression and the Second World War formed a great divide in human history. On the far side of this divide were the nineteenth and early twentieth centuries, with their urban factory culture, replete with grasping industrialists and deprived immigrant workers. On that side also were slums, assembly lines turning out consumer durables such as cars and refrigerators, and unions, which brought necessary equity to a workplace in which the values of the nineteenth century still held sway.

On the near side, where these men stood, were suburbs, a more homogeneous society, and plants in which the motif was harmony because their owners, managers, and workers were essentially the same people. This view was consonant with the changes in American life that followed the Second World War. It reflected the assimilation into the American mainstream of the second and third generations of immigrant offspring and the generally higher standard of living that the economic boom of the fifties and sixties had bestowed on the nation.[29]

These employers inhabited a world in which innovation, hard work, and ever-greater productivity were constants. They brought as social baggage to this new world not the class consciousness nor the Social Darwinism of the nineteenth-century industrialists, but small-town Puritanism, with its emphasis on hard work, equality, and community. They sought on the plant floor not the extraction of energy from the hostile and alien forces of labor, but cooperative effort from their fellow employees, who were seen almost as family members. Labor in the factories of the high-technology entrepreneurs was not to be objectified as in the earlier era, but humanized. It was to be nurtured, not exploited; new ways to keep its loyalty and spur gains in productivity were to be experimented with, as befitted people who constantly sought innovation in the technological sphere.[30] Even the titles of high-technology management reflected the change in attitude toward workers: they were no longer "personnel managers" but "human resource" directors. In such a setting, entrepreneurs saw unions as intruders.

The pioneers of Washington County's high-technology industry, Vollum and Murdock, took great care to build into the atmosphere at Tektronix a familial feeling.[31] From the firm's earliest days Murdock, who was largely responsible for labor-management relations in the company's formative years, had expressed a strong distaste for unions and a preference for a small plant in which everyone was known on a first-name basis and all participated in company decisions.[32] Even after the company far surpassed the threshold (one hundred workers) at which that seemed likely to work, what had come to be called the Tektronix culture was still pursued.[33]

As we have seen, that culture was based on high productivity, quality, and efficiency in every phase of the business through mutually shared goal setting, teamwork, and risk taking. The main way in which these characteristics were nurtured was profit sharing. Installed in 1949, profit sharing totaled 25 percent of Tektronix's pretax profits in the first year, paid out as a proportion of each employee's salary.[34] Because Murdock and Vollum had modest financial goals for themselves, they were willing to share Tektronix's largess with their employees: with no regrets they watched profit sharing reach 62 percent of pretax profits late in the fifties.[35]

So far as management was concerned, the benefits of profit sharing were highly practical. Many of the usual problems of big companies, such as waste, theft, malingering, and low morale, were largely eliminated at Tektronix because, in a sense, workers would be stealing from themselves. At the same time, creative, ambitious people were attracted to the company, enticed by the possibility of remuneration far beyond a stated salary. And, perhaps most important, the need for union intervention in company affairs seemed to be obviated.

Another device of similar impact in the workplace has been the "quality circle." Called by a variety of names throughout the industry, the quality circle is characterized by periodic meetings of workers at all levels within a company's divisions or working groups. There, grievances and suggestions about various problems, ranging from personnel matters to product improvement, are aired. Frequently, solutions to the problems presented can be effected by those present. At Tektronix, ESI, Intel, and other companies throughout Washington County, the quality circle has been a surrogate collective bargaining agent, apparently diminishing the need for a real one within the plant walls.

Another reason for the inability of unions to convert the county's electronics labor force has been a twenty-year-long movement in both the federal and state legal systems to extend the protection of the law to employees in many cases in which job security, job status, discrimination, or privacy in the workplace are at issue. Thus in 1974, in *Monge* v. *Beebe Rubber*, a New Hampshire case, the court ruled that a dismissal motivated by bad faith (in this instance, sexual harassment) was ground for an employee to bring suit; in *Toussaint* v. *Blue Cross & Blue Shield*, a 1980 Michigan case, the court determined that employees could use printed company statements and proof of oral assurances of job security to establish a contractual right to be fired only for just cause.[36]

In Oregon, these landmark cases have actually been surpassed. For example, a 1984 decision, *Holien* v. *Sears Roebuck*, went further than *Monge* in that it widened the grounds for awarding damages in such actions.[37] Likewise, *Sabin* v. *Willamette Western Corporation* and

Yartzoff v. *Democrat Herald Publishing Company* strengthened the hand of plaintiffs in circumstances similar to those of the Toussaint proceeding.[38]

In effect, these and other cases successfully provided protections for workers that once might have resulted only from union-instigated grievance procedures. At the same time, human resource professionals have observed growing limitations on management's discretion to dismiss employees, leading perhaps to a greater respect for their workers' positions and power.[39] This, in turn, has no doubt suggested to workers in high technology—and for that matter elsewhere in American industry—that unions are superfluous to their needs.

Naturally, organized labor, particularly the Communications Workers of America, has taken a different view. With only one Washington County high-tech employer organized and an extremely expensive effort required to get more, the union has been hard pressed to get its views heard by workers in the industry. Nevertheless, it has not given up and has drawn for workers a somewhat grimmer portrait of the industry than the one to which they are accustomed. For instance, while the average hourly wage of high-technology workers (50 cents an hour higher than in forest products, according to the American Electronics Association, or about $6.00 to $7.50) is not bad, the union has claimed that it is less than a wage negotiated at the bargaining table would be.[40] In 1986, the union projected that a $12.00 hourly wage would result from collective bargaining.[41] Another of the union's assertions is that the current average wage is not really adequate to enable workers to purchase the customary perquisites of the American middle class, such as a home.[42] Nor, it is said, can workers even pay taxes or "keep a child in school."[43] The union has also said that the benefits packages of Washington County high-technology workers could be vastly improved.[44]

In addition, organizers have noted that, besides increased wages and benefits, collective bargaining would gain for workers strong layoff clauses, which would include sufficient notice of impending work-force reductions and the elimination of tenure as a criterion in selecting workers for layoff.[45] In light of the sweeping layoffs at both Intel and Tektronix over the first six years of the eighties, this might prove to be an incentive for workers to invite the Communications Workers of America to represent them in the future. In late October 1985, on the cusp of the electronics slump, at least "two quiet" organizing drives were undertaken at electronics companies in Washington County, according to Linda Rasmussen, president of Portland's Local 9201 of the Communications Workers.[46]

At that date, however, the outlook for union success still seemed nebulous. For example, Douglas Strain, ESI's retired chairman, felt confident that the older companies would maintain good relations with their employees—even in the face of the longer hours and wage rollbacks demanded by hard times.[47] Such companies were experienced at meeting employee unrest. "But take a company like Mentor," he said. "It's never known anything but a super boom. They don't know how far they'll have to adjust [to declining sales]."[48] The implication was that the new generation of executives, the young entrepreneurs of high technology, experienced in growing a company but not in managing a slump, might antagonize their work force through ruthlessness born of panic, thus offering the union an entering wedge.

In any event, the union has claimed that, historically, organization of electronics companies has not hurt employers vis-a-vis cheaper competition.[49] The Communications Workers of America point to their one union shop in Washington County as proof of this. When the company faced severe financial difficulties, the union worked with the employer to save his firm (which the union refuses to name). Its help included "give-backs," coupled with safeguards for recapturing lost wages and benefits in the event the company returned to profitability.[50]

In sum, the union has maintained that despite innovations, which ostensibly give workers participation in company decision making, the Washington County high-technology industry is not vastly different from other, less progressive ones.[51] Workers are treated much the same and must battle for, and protect, their rights, their wages, and their jobs. Whether this view will persuade workers of the efficacy of collective bargaining remains to be seen.

THE INDUSTRY AND ITS ENGINEERS

In addition to attempts to organize workers, there has been talk of organizing engineers in electronics companies. Neither in Washington County nor elsewhere, however, has such an effort been launched. Moreover, it would probably be unsuccessful, because electrical engineers represent the lifeblood of the industry and have been in chronically short supply throughout the eighties. They have no need to strike for higher wages or better conditions, because their services are already at a premium.

"Engineers are especially hard to organize," one union official has admitted.[52] "They don't feel they would benefit from a union, unless they're employed by the government, by aerospace companies, or by AT&T, where the work situation is more bureaucratic. In those companies, they may feel they've lost the ability to advance on the basis of their talents."[53] On the other hand, engineers in young companies have usually been too preoccupied with the challenge of their projects to bother organizing.

Certainly engineers have felt the insecurity associated with layoffs in the industry, but they appear never to have considered the possibility of inviting a labor union to solve the problem.[54] "There's a psychological perception that they're the wave of the future," one scholar of organized labor has commented.[55] Largely self-motivated and engrossed in their work, they feel it is "the responsibility of management to deal with employees' concerns."[56]

Employers have necessarily recognized that their engineers' happiness and loyalty are of primary importance to their companies' success. The nation's supply of electrical, computer, and software engineers has been so limited that the recognized shortage has become a matter of intense concern. In relatively isolated, sparsely populated areas such as Washington County, this is especially true. The combined graduates of Northwest college and university engineering programs are not numerous enough to make significant inroads on the demand, even of one large company like Intel.[57] Thus, Intel and Tektronix must compete nationally for new graduates by sending waves of recruiters to the placement centers of the country's colleges and universities and by offering internships to undergraduates in the hope that they will remain loyal to the company upon receiving a degree. Companies actually prefer recent graduates to established professionals, because it is cheaper to recruit a new graduate than to pay the relocation costs of an experienced professional, which can be as much as fifty thousand dollars from California.[58]

The growth profile for electrical and electronics engineers in the Portland area for the period 1982 to 1990 is startling. In 1983 there were 3,340 such engineers and in 1985, 3,930. In 1987 there were projected to be 4,320 and by 1990, 4,910.[59] This represents job growth of approximately 47 percent, even in the face of the major reverses of the eighties. At least one executive has warned that the companies' insatiable demand for engineers has begun a dangerous trend of drawing college professors out of the classrooms and into the plants, where they can get much higher compensation. As Tom Bruggere of Mentor Graphics put it, "We are eating our seed corn."[60]

However that may be, it is clear that Oregon alone cannot meet the needs of the industry in Washington County, not to mention the state, as far as engineers are concerned. For the entire state in the period from 1983 to 1987, only 31.4 percent of the new employees coming directly from college were projected to be from Oregon. Nearly 70 percent were to come from the rest of the United States and elsewhere, while more than 40 percent of new employees from sources other than the colleges would be from outside Oregon.[62]

In this connection, there has been one ray of hope. The American Electronics Association released figures in 1984 showing that Oregon's turnover rate for engineers was significantly lower than the national average: 12.1 percent, compared with 17 percent.[63]

If the scarcity of talent were not enough to make employers pay attention to the needs of their engineers, the industry's greatest imperative—innovation—would demand that their requirements be met. Most high-technology firms in Washington County, as in Silicon Valley and along Route 128, have striven to provide a setting that is both nurturing and challenging to the engineering staff. At Intel, management always has set great store by youth. Robert Noyce's view was that the "great flashes come from the young."[64] His own invention, the integrated circuit, and that of Ted Hoff, the microprocessor, were achieved when both men were relatively young. Squadrons of young professionals at this firm have therefore been given their heads to imagine and invent.[65]

To stimulate this impulse, the founders deliberately subverted the protocols of traditional corporate America. For the most part, rank and privilege were abolished. Everyone worked together in large rooms at desks of equal size, separated from each other only by occasional partitions. "Strategic business segments," the Intel equivalent of corporate divisions, were set up, with middle managers given more latitude than vice presidents in more traditional companies. Meetings were called frequently, and young engineers were expected not only to bring up their own new ideas but to challenge those of the legendary Noyce, Gordon Moore, and Andrew Grove.[66]

The atmosphere at Intel, as well as firms in the Silicon Valley, was volatile. Intellectual clashes were a common feature of the corporate meetings, and extended days and nights of R & D work the norm. "They yell a lot at each other," was the way one book containing sketches of the nation's best employers put it.[67]

Intel has consistently had a high turnover rate, but the long-term success of its products and the challenge of its environment argue that its style has been successful for the company generally and for many of its employees. For example, in 1981, as an incentive in addition to its stock-option program, it introduced a program of sabbaticals of from eight weeks to six months. Although there was concern that large numbers of employees would take advantage of the leave and not return, only 6 of the 302 who took sabbaticals in the first year failed to resume their positions.[68]

A world away from the tensions of Intel is the environment of Tektronix. Although its egalitarian atmosphere matches Intel's, tension has never been associated with employment at the older firm. Instead there is a paternalistic atmosphere, which some observers believe has cost the company its edge in recent years.[69] Inertia and stagnation have set in, they say, and competitive drive has been lost.

The new companies in Washington County have observed the

patterns of the older ones and tried to establish personalities that combine their best attributes in order to stimulate productivity and innovation. At Mentor, for instance, it is hoped that the formation of divisions and the push to include more and more people in making decisions will increasingly concentrate authority in the hands of line engineers.[70] At the same time, there is a constant effort to winnow out the bottom 10 percent of producers in the company. Although the atmosphere is less charged than that of Intel, it is not as relaxed and placid as that of Tektronix.

Employers worry that their technical professionals will leave and form spin-off companies. As we have seen, that is how new firms customarily start. Spinning off represents a large problem—not simply because of the obvious brain drain but because of the high recruitment costs, increased competition, and R & D setbacks resulting from the defection of key individuals.

According to venture capitalist Ralph Shaw, engineers are likely to become restless, not when business is bad or even especially good but when companies get smug about profit margins and cease funding research and development at high levels, thus curtailing the primary work of engineers habituated to the excitement of experimentation.[71] Companies such as Tektronix have attempted to fight this tendency by forming "intrapreneurship" (internal) ventures and by keeping the level of research and development expenditures high.

With the expansion of the electronics industry in the county, the rate of job turnover should rise. Ironically, this would be a positive indication that enough firms were doing business in the area to make engineers feel that they could migrate to greener pastures in the vicinity with some impunity.[72] This could well attract more experienced engineers, in turn, making the industry more innovative. More innovation should lead to greater profitability for companies, and with that should come expansion. If the industry expands greatly it may, in time, make realities of the complex myths associated with the county's high-technology labor force.

10

The Balance Sheet

As American industries go, high technology is new, and its effects on Oregon, especially on Washington County, are varied. It is one of the state's largest employers, and Tektronix is its largest company. But the industry has made many contributions besides creating jobs and products. Just as high-technology firms spin off new companies, so high technology itself spins off community benefits. And, just as the industry has had its dark moments, it has sometimes had negative effects on the larger community.

In 1951, when Tektronix built a new factory at what is now the intersection of Barnes Road and Highway 217, its founders did not realize that their decision would catalyze a boom in the county's industrial real estate. In choosing the relatively undeveloped, uncrowded outskirts of Beaverton rather than Gresham, Jack Murdock and Howard Vollum predestined Washington County to be the seat of high technology in Oregon.

The marriage of Washington County and high technology has been assisted by matchmakers—commercial and industrial realtors, investors, landowners, local governments, and various booster groups—who have made vigorous efforts to present Washington County's real estate as the most attractive available.

The most prominent of these entities has been the Sunset Corridor

Association, an organization that is largely the result of the efforts of one man, Wayne Atteberry. Atteberry was vice president of real-estate finance for Standard Insurance Company of Portland and the owner of eight hundred acres in Washington County. His efforts began when his firm and two other landholders, Sequent and Quadrant (the development arm of Weyerhaeuser), formed a local-improvement district to upgrade Cornell Road, the arterial servicing their properties.[1] Although the three were competitors, they all needed a good road if they were to develop their land.[2]

Once the efficacy of the concept was recognized, Atteberry worked to bring others into the group. An organization soon evolved encompassing most individuals and organizations with a stake in developing the area known as the Sunset Corridor. By 1984 fifteen major property owners had joined the group.[3] Atteberry, whose company owned about 40 percent of the land, became its president, and in 1985 his wife, Betty, an experienced political operative, was named executive director.[4]

The association's goals were to create "an area marketing concept," as Atteberry labeled the strategy, and to respond to the common problems the property owners faced.[5] There was much to be done to improve the area's potential, especially the lobbying of government bodies to widen roads, improve dangerous intersections, provide traffic lights, and standardize signs.[6] In addition, the association could secure "environmental amenities," which, in the words of a Washington County planner, would unify the area "with an aesthetic vision."[7]

Among the organization's developments are the 160-acre Oregon Graduate Center's Science Park, the Wachovia (North Carolina) Bank and Trust's 112-acre Cornell Oaks Corporate Center, the Sea-Port Industrial Group's 197-acre West Union Park, and Riviera Motors' 128-acre Five Oaks Industrial Park.[8] They are sites desired by electronics concerns, but they represent what one observer, writing in *Oregon Business*, called a "nearly perfect mix of land uses."[9]

The Sunset Corridor Association's success is confirmed by the price of its land. By 1986, farmland that sold for $3,000 an acre in the early seventies was priced from $60,000 to $80,000 an acre for serviced campus-style parcels.[10] Among the buyers have been Epson, Fujitsu, and NEC. Although in 1986 land sales and leasing slowed because of the worldwide slump in the electronics industry, the membership remained optimistic and planned for a rebound.[11]

Three linked factors explain the success of the association and other lesser known land developers, such as the Tualatin Valley Development Corporation. In the early eighties the crisis in Oregon's wood-products and fishing industries made it clear that government and business would have to cooperate to move into a more highly diversified economy. The Sunset Corridor Association's successes had made government eager to explore a partnership. Planners estimated that between 1984 and 1994 the corridor might grow from a population of five thousand and a job total of two thousand to a population and job total of thirty thousand each.[12] This type of government optimism helps to explain the ease with which the developers' needs were met.

The relationship between government and business, however, is not without its ironies. For one thing, some professional planners believe that the state's land-use watchdog, the Land Conservation and Development Commission, has had a salutary effect on Washington County's development. A favorite scapegoat of many in the development community since 1973, the LCDC has promulgated nineteen planning goals, to which local governments must adhere. Although these goals were designed to strike a balance between economic exploitation and environment conservation, it has frequently been charged that they have retarded or prevented development.[13] However, one Washington County planner noted that "LCDC forced us to do a good job. We learned a lot about industrial parks that we did not know."[14] A developer conceded that LCDC goals allowed the "large users" to "protect themselves from any development around them... [of] less than the highest quality."[15]

Another irony is that Oregon's unattractive tax structure has not really harmed development. Although high property taxes are unalleviated by a sales tax, this state of affairs has not made it as hard as one might expect for the county to compete with Seattle or Los Angeles in the race to attract high-technology firms. In 1984 Portland property taxes were, at $24 per $1,000 of assessed valuation, double those of the other two cities.[16] But the price of Portland land, at $3 per square foot, was half the cost of Seattle's and one-quarter that of Los Angeles.[17] At the same time, Portland's construction costs reflected comparable advantages over the other two cities.[18]

With one notable exception, the future seems equally bright. By the end of the eighties, the Sunset Corridor Association may run out of developable land unless the urban growth boundary, adopted in 1979, can be extended. Although the boundary was supposed to accommodate growth in the Sunset Corridor until the turn of the century, unanticipated development during the eighties almost exhausted the available large-scale sites.[19] A 1985 report by an economics-consulting firm on the subject stated that

> the reduction of variety in land choice constitutes a constraint on the ability of the Sunset Corridor and Portland to sustain growth in high tech industries. The typical purchasers of large sites are major, immigrant firms. These firms are important engines of growth by virtue of the size of their employment needs, the value-added in production which is dispersed within the local economy, [and]... the increase in the local tax base.[20]

Naturally, the Sunset Corridor Association would like to expand the urban-growth boundary to address this problem. Considering that only thirty-five years have elapsed since Murdock and Vollum made their crucial decision, it is a remarkable and not unpleasant dilemma.

PHILANTHROPY

Another obvious community benefit of high-tech companies has been their philanthropy. In 1952, Tektronix created the first major high-technology foundation in Oregon, and since that time its gifts have been of enormous benefit to Washington County. Among the recipients have been St. Mary's Home for Boys, St. Vincent Hospital, the

Tualatin Valley Mental Health Clinic, the Washington County Council on Aging, and the Washington County Historical Society's new museum fund. Tektronix also gives scholarships to the children of its employees and to science and mathematics students in the county. In addition, between 1960 and 1985 Tektronix and its employees gave approximately $19 million to the United Fund campaign. In recognition of these and other community services, the Beaverton Area Chamber of Commerce gave the firm its 1986 J. Arthur Young award for contributing to the quality of life in the area. It was the first time the award had been given to an institution rather than an individual.[21]

The second great philanthropic by-product of Washington County's high technology was the M.J. Murdock Charitable Trust, which was established in the will of that Tektronix cofounder and formed in 1975. It immediately became the largest private foundation in the Pacific Northwest, with first-year appraised assets valued at $91.6 million. Although Murdock's will set general guidelines for the trust, its trustees soon developed a grant program that spanned an array of contributions to human welfare in seven categories: education, scientific research, community human development, human development for the handicapped, health and medicine, arts and culture, and public affairs. In its first decade the trust made 960 grants, totaling $74.9 million, to 471 organizations. Ninety-four percent of the money went to institutions in the Pacific Northwest. Among those in Washington County that benefited were the Oregon Graduate Center, Pacific University, the Oregon Regional Primate Research Center, and the Washington County Historical Society. Through careful management of the trust during its first ten years, total assets rose from $91 million to $169.1 million, while the overall investment return was an impressive $152.1 million.[22]

Besides foundations, many individuals connected with the Washington County high-technology industry have given back much to the Portland metropolitan community. Together or separately, C. Norman Winningstad of Floating Point Systems and his wife, Dolores, have involved themselves in many cultural endeavors. For example, they have donated one of the theaters in the Portland Performing Arts

Center; twice underwritten the costs of lectures by Jane Goodall at the Washington Park Zoo; funded an exhibit on holography at the Oregon Museum of Science and Industry; paid for the services of a live orchestra at a performance of the Pacific Ballet Theatre; donated money for the construction of the Washington County Museum; contributed the entire expenses of a concert by the Oregon Symphony Orchestra; and sponsored programs like "Sesame Street" on Oregon Public Broadcasting. Their generosity reflects Dolores Winningstad's philosophy: "When a community has been good to you, you owe something in return."[23]

Howard Vollum also donated to many civic and cultural institutions. He was a founder of the Oregon Graduate Center and a major donor to Reed College, and with his wife, Jean, contributed to many other causes and institutions, including Oregon School of Arts and Crafts, the Portland Art Museum, Catlin Gabel School, the Oregon Museum of Science and Industry, and St. Vincent Hospital.[24]

WASTES

No one in Washington County's high-technology industry would claim that its contributions to the community were without price. Any large industry makes an impact on its social and natural surroundings and sometimes causes alarm or fear rather than respect or gratitude. Certainly one cause of alarm about American high technology in the eighties is its real or potential effect on the natural environment.

The sixties were marked by growing environmental consciousness that translated into a political force of some importance at both national and state levels. In 1963 Congress passed the Clean Air Act, in 1965 the Water Quality Act, and, over the years, other legislation to cleanse or protect the environment, while in 1969 the Oregon legislature created a Department of Environmental Quality. In the seventies the environmental movement helped bring about passage of the National Environmental Policy Act and the Federal Environmental Pesticide Control Act and creation of the Environmental Protection Agency by executive reorganization. In 1980 Congress passed the Resource Conservation and Recovery Act to codify the toxic-waste regulations of the past nine years.

These regulations—aimed primarily at industry—seemed imperative because of past violations of the public trust. The new attitude was that never again would the public take on faith the claims of corporations—particularly chemical corporations—concerning the safety of their operations. In the seventies nuclear energy and methods of toxic-waste disposal, in particular, came under attack because of the potential for leaks. In retrospect these attacks appear justified after the spectacular disasters that have taken place at Love Canal, Three Mile Island, and Times Beach, among others.

The equivalent disaster for the high-technology industry was in south San Jose. In December 1981, it was found that a storage tank at the Fairchild Semiconductor plant in that city was leaking a dangerous chemical, trichloroethane, into the earth and ultimately into an adjacent well. Fairchild notified the water company, which closed the well the next day, but neither company disclosed the discovery to the public for fifty days. When the news broke, investigations of other disposal tanks revealed eleven leaking tanks in the Silicon Valley. Several of the largest firms in the region were implicated, and many lawsuits were filed against them for negligence. Plaintiffs charged that the leaking chemicals had poisoned water supplies, which in turn had caused birth defects, stillbirths, and miscarriages. The incidents drew national interest and led to continuing publicity. Among other things that came to light as a result were earlier studies suggesting that electronics employees may have suffered work-related cases of poisoning, respiratory disease, chemical sensitization, hypertension, radiation exposure, reproduction problems, and cancer.[25]

In Washington County, environmental problems involving high technology have also come to light in recent years, although compared with those receiving national attention, they have been minor. One was discovered in February 1985, when Tektronix was conforming to federal law by testing soil and groundwater and by researching past hazardous-waste-handling practices. In the process company scientists found contaminated water on its property below two sites that had been used in the early sixties for waste-storage ponds. The water contained potentially dangerous levels of trichloroethylene, an all-purpose industrial solvent that had been widely used in the United States for thirty years. Three months later, in May, the company found that another lagoon was definitely toxic. The company had maintained three treatment ponds at the site between 1963 and 1979, where it had placed the sludge from its detoxified manufacturing wastes. (During these years the material had not been considered hazardous waste by the state's Department of Environmental Quality and the company's operations had been perfectly legal. Since 1979 Tektronix's hazardous wastes—like those of all Oregon corporations—have been taken to a disposal site in Arlington.) Tek responded to the discovery of its toxic-waste lagoons with continued testing and cooperation with the Department of Environmental Quality and a pledge to clean up the contaminated areas.[26]

There is no question that the use of chemicals continues to be a potential danger of Washington County's high-technology industry. In 1980 the Department of Environmental Quality reported that Washington County was producing 23 percent of the state's hazardous waste, most of which came from high-technology plants. By 1985 the DEQ listed fifty-two known businesses in Washington County "where significant levels of toxic wastes are produced as part of routine commercial operations"—most of which were in the electronics industry. Workers in the industry, as well as other residents, were concerned about the potential dangers and the number of dangerous chemicals used in the various corporations' manufacturing processes.[27]

Responsibility for protecting against these dangers is shared by government and industry. The Accident Prevention Division of the Workers' Compensation Department enforces state rules for the safety of workers, including those who handle toxic and hazardous materials. Its powers were enhanced by a 1985 Oregon statute requiring companies to post notices of dangerous materials used in their plants. The Department of Environmental Quality licenses and inspects manufacturers of toxic wastes; develops rules and standards for handling the wastes; responds to complaints from the public; investigates spills; oversees industrial-waste treatment, detoxification, and disposal; and supervises cleanup of spills or contaminated areas. Industry is responsible for following federal and local regulations that involve, among other requirements, employee training in the safe use of chemicals, ventilation of the workplace, and detailed record keeping from the arrival of chemicals to their disposal. Remedial measures are also under way.[28]

In Washington County, the Oregon Graduate Center was involved in two research projects dealing with industrial wastes. One, under the sponsorship of Michael Gold, attempted to discover how microorganisms could be used to eliminate toxic wastes. Another, headed by James Pankow, was an investigation of groundwater: how it spreads and how chemicals spread through it.[29]

WOMEN IN HIGH TECHNOLOGY

Women employees have always been indispensable to the high technology industry, but they have not shared equally with men in its benefits. Much of the blue-collar work in electronics firms—nationally and in Washington County—is done by women, who in 1983 constituted 44.2 percent of the electronics production-line employees in the United States. One reason Tektronix moved to Washington County during the fifties was the existence of a potential labor pool of women, who have remained important as production-line workers at Tek and at newer firms since then. The story for women at middle- and upper-management levels, however, has been different. In 1983 women made up only 11.3 percent of employees at those ranks nationally; among Washington County firms their participation was also low. In 1985 women made up only 1 percent of the officers and members of boards of directors of the county's locally based public high-tech corporations (ESI, Flight Dynamics, Floating Point, Mentor, and Tektronix). At the Oregon Graduate Center in the same year women numbered a mere 5 percent of the full-time faculty (two out of thirty-seven).[30]

The reasons for the small number of women in high technology's middle and upper echelons are diverse. One explanation is that girls and women have traditionally eschewed studying science and engineering or have been discouraged from studying it. Another is that electronics either is, or is perceived to be, more abstract than other professions and therefore might not seem to appeal to women. A third is gender prejudice in hiring and promotion. Although the explanations for this avoidance differ, a 1984 National Science Foundation study reveals that one in four scientists is a woman, while one in twenty is an engineer, and that women are less likely to go into electrical engineering than any other engineering. Even when women do obtain management jobs, they begin at a lower pay rate than their male equals and never catch up to them. Their rates of promotion are also slower. The responsibility of government to close these gaps derives from the Civil Rights Act of 1964, administered by the Equal Employment Opportunity Commission, which prohibits discrimination on the basis of race, color, religion, sex, or national origin. An Executive Order in 1965 required that every government contract contain an equal employment-affirmative action clause. Since 1972 every company employing twenty-five or more persons has been required to file a report with the commission. The record of high-technology firms in Washington County shows that they have met the government standards for affirmative action. Any further improvements will depend on the companies' ingenuity in finding ways to bring more women into their highest management positions.[31]

NEW JOBS OUTSIDE THE INDUSTRY

The most important contribution of the high-technology industry, in the mind of the average citizen, is the creation of jobs. Washington County's high-tech firms have generated much employment, not only within the industry, but in a variety of allied enterprises. In advertising and public relations, for example, one of the most successful new local companies is Karakas, VanSickle, and Ouellette (KVO), founded in 1983. Steve Karakas and Sharon VanSickle had worked at Tektronix, while Pierre Ouellette had done contract work for that firm. KVO's first advertising client was Mentor Graphics, and Tektronix was an early user of its advertising and public-relations services. (Steve Karakas refers to Tektronix as KVO's "venture capital" firm.) By February 1986 KVO had a client list of fifteen high-technology firms, many of which were in Washington County.[32] Another new public-relations firm begun in 1983 is The Waggener Group, founded by Melissa Waggener and Julie McHenry, who had worked for Regis McKenna (the noted electronics public-relations firm based in California) and then moved to Tektronix for a time before deciding to strike out on their own. The firm numbers ATEQ Corporation and Sequent Computer Systems among its clients. In addition to these firms, which owe their existence to local high technology, there are other success stories. Regis McKenna opened a Portland office in 1979 to serve local clients.[33] An older Portland firm, Young and Roehr (founded in 1968 by Jeremy H. Young and Frank Roehr), was able to add to the forest-products firms that were its first clients the Tektronix Information Display Group and other high-technology firms, including Floating Point Systems.[34]

The legal profession has also gained from high technology. Some of the older Portland firms were somewhat slow in adapting to the new enterprises, being more experienced in forest products and other traditional industries, and some newer high-technology firms first found legal counsel in a Seattle firm, Perkins, Coie, Stone, Olsen and Williams, which opened a Portland office in 1983. In time, however, local firms were attracted to high technology. For example, in 1984 Oregon's largest law firm, Stoel, Rives, Boley, Fraser and Wyse, opened an office in Washington County, primarily for the industry's concerns.[35]

Although in the Silicon Valley, Route 128, and other high technology centers there has been litigation concerning trade secrets, patent rights, and employee recruitment, law firms in the Washington County industry have been concerned with more prosaic, though equally important, matters. Perkins, Coie, for example, was active in representing venture-capital firms that were investing in local high-technology firms. Among the firms it assisted were Metheus-CV, Northwest Instruments Systems, Planar Systems, and Sequent.[36]

VENTURE CAPITAL

It is said that the secretaries in a local high tech firm were able to pinpoint the moment at which the founders' quest for venture capital was successful: it happened the day their talk turned from research on the company's embryonic product line to the purchase of BMWs, condominiums in Maui, and elaborate sound systems. This anecdote may be apocryphal, but it does capture the way in which financial backing can dramatically transform a promising shoestring operation into something real and lucrative.

While new firms of every description have always sought capital from various sources, by the mid-eighties the term "venture capital" had become almost synonymous with the placement of professional investors' funds in concerns related to technology. There are many reasons for this. First, alterations in the U.S. tax code, including reductions in capital-gains taxes, attracted capital to new, potentially productive fields.[37] High-technology start-ups are appealing because they are known to produce approximately four times as many innovations per R & D dollar as medium-sized firms and twenty-four times as many as large firms.[38]

Second, for most of the eighties and especially in 1983, investors showed great acceptance of initial public offerings of high technology stock, which promises liquidity.[39] (In 1984 Mentor Graphics—then three years old—was reputed to have made the most successful public offering of the year, raising nearly $50 million.)[40]

Third, owing to the changed structure of the American financial markets since the Second World War, large amounts of investment funding had been available for nontraditional investment. Much of this funding was channeled to venture firms looking for new companies to back.

Some wealthy individuals or families were attracted to high tech for their investments, but the largest amounts of venture capital were generated by pension funds, banks, and corporations. In 1984 pension-fund moneys available for investment nationwide reached an astronomical $1 trillion.[41] The situation was different from pre-Depression days, the time of greatest investment activity before the Second World War. At that time, 5 percent of the population controlled 33.5 percent of the nation's wealth, and investments were made by individuals and orchestrated by a few conservative brokerages and investment banks in Boston and New York.[42]

Over the years since 1945, however, this pattern had changed, until by the seventies the largest investors were institutions, such as pension funds. The managers of institutional investment funds must maximize returns while minimizing risk, leading them to diversify portfolios from the tried-and-true Dow Jones blue-chip corporations to new, higher yielding possibilities such as high technology. The highly publicized, apparently phenomenal success of high-tech start-ups such as Apple, Intel, Mentor Graphics, and Sequent made electronics companies unusually alluring to capital managers.

Because these companies require large infusions of capital while research and development proceeds and no money is being earned, investors characteristically have a large equity in the business and exercise an element of control, often in the form of one or more board seats. Investment professionals like such a situation because it affords them close oversight of their funds and, because the high-technology market has been in a continuous state of revolution since the mid-fifties, innovative new products (the *raison d'etre* for start-ups) can readily find a lucrative market, sometimes bringing improbable returns on investments in only three years.[43]

A central irony of the quest for high-tech venture capital is that for neither entrepreneurs nor investors is technological innovation a primary consideration in starting a company or investing in one. Only rarely have interviews with corporate founders yielded the admission that their main goal was to push back the frontiers of technology. Although pioneers such as Robert Noyce and Gordon Moore of Intel were scientists of great accomplishment, neither was driven by a vision of advances in research but, rather, they formed the company because they were disgruntled with the bureaucracy of Fairchild Semiconductor. Tom Bruggere was not first and foremost a scientist, but a manager at Tektronix, when he and his partners left to found Mentor Graphics, and his primary motive in the undertaking was to run a major publicly held corporation.[44] Casey Powell, the former Intel engineer who cofounded Sequent Computer Systems, said he and his colleagues had no product in mind when they started their corporation.[45]

In any case, for venture capitalists the keys to prudent electronics investing have less to do with trailblazing products than with the founders' characteristics and demonstrated business acumen. Most financiers have sought to invest in the enterprises of traditional, cautious entrepreneurs because they are aware that successful firms in the field have survived by practicing the same business methods as successful firms in other fields. Moreover, financiers know that successful high-technology start-ups have displayed certain characteristics common to traditional business practice. These include multiple founders, each specialized in one of the critical areas of any company—research and development, marketing, finance, and manufacturing; a chief financial officer who has been a professional in the field; a chief of marketing who is oriented to research as well as sales; founders who are experienced in firms with at least five hundred employees; and a record of heavy spending for research and development.[46]

These and other critical data are revealed to the venture capitalist studying the business plan of the potential start-up. The plan—a comprehensive, well-documented projection of the new concern's structure, goals, and financial needs—typically consists of a concise description of the company's history and concepts; resumes, references, and other relevant information on the principals; a listing of the firm's owners (those individuals or institutions holding more than 10 percent of its stock); a full description of the kind of business in which the firm is to engage, including products, processes, proprietary technology, and patents; elaboration of the current size and potential growth rate of the market for the product; a description of the competition; a financial profile, including past, present, and future financial data, projected for five years; a statement of current and future financial needs; and a statement of other needs, such as advertising, that the firm is likely to have.[47]

Ralph Shaw, principal of Shaw Venture Partners, a Portland firm, has said that in addition to the written presentation, the financier screens entrepreneurs through discussions that reveal personalities, temperaments, imagination, and other traits.[48] These discussions also explore further the projections of entrepreneurs, who have appeared in his offices at various stages in their companies' development—80 percent with a business plan in hand, 20 percent with only a concept, 15 percent with a product.[49] Those most likely to succeed can fully justify their forecasts, have clearly identified marketing problems and their solutions, and have secured reliable suppliers.[50] In the end, however, the business plan is only a product of the management team's intelligence and imagination; therefore, most investors say they are really investing their funds in the management team itself.

Ultimately, venture capitalists must be able to project a return on their original investment in multiples of it and have a sense that the firm will grow to a minimum size within five years (the standard time period for business projection).[51] If everything is in order, a deal is usually struck. However, the necessary elements are seldom all in order. Since its founding in 1983, Shaw Venture Partners has scrutinized between 750 and 800 business plans but invested in only 13.[52] Nor, for all the scrutiny lavished on start-ups, is there any guarantee that the select few eventually subsidized will return a profit on the investment. For example, dilithium Press, a software publisher in Beaverton and one of Shaw's thirteen investments, was forced into bankruptcy because of enormous debt burden compounded by a shift in demand for its computer "how-to" books."[3] Star Technologies, a high-performance computer manufacturer and another of the firm's investments, also closed its doors because of financial difficulties, as indicated by the fact that its first stock offering, which traded at $9 a share in 1985, had sunk to $3 a share in 1986.[54] On the other hand, such companies as Integrated Measurement Systems and Sequent Computer have proved to be excellent investments for Shaw.

Although venture capitalists of the kind described have been by far the largest lenders of sums and the most reliable sources for high-technology entrepreneurs in Washington County and elsewhere in the nation, there are other methods of funding and other kinds of lenders, which in recent years have become active and important financial alternatives available to entrepreneurs. Among them are the more adventuresome banks. In the lead in Washington County has been Far West Federal Bank, a thrift institution, which—like others of its kind—crossed the line from offering high-interest savings and car and mortgage loans to competing in a variety of ways with commercial banks after deregulation of the banking industry was mandated in 1982. While Far West has not financed many electronics firms, it has worked out loan packages for a few, such as Audio Precision, a firm founded in 1984 by four former Tektronix employees.[55]

The bank's approach, known as "mezzanine funding," differs significantly from capitalization by venture firms. In a mezzanine deal (so called because it may occur just after a company is out of the start-up phase) the bank gives $250,000 to $500,000 in funds, along with additional moneys in the form of traditional bank loans.[56] The funds carry an agreement for a preferred return of 30 to 50 percent, which might be in the form of a warrant to buy the company's stock at a guaranteed price as well as the right to sell it back to the company whenever appropriate.[57] The conventional loans are given at the prevailing interest rates.[58] In such an arrangement, the bank's risk is limited because, unlike venture-capital firms, it has no ownership in the young company.

By 1985, about two years after the program was initiated, Far West had lent nearly $10 million in mezzanine funds.[59] The program, meanwhile, had come to be regarded as unique, occupying ground between "early-stage equity financing," as one observer put it, "and pure debt bank loans."[60]

Another type of funding that has helped a few high-technology companies in Washington County is known as "corporate partnering," which had a brief vogue in the sixties, until it failed and then reappeared—with some differences—in the eighties.[61] Corporate partnering disappeared after some large firms acquired small companies at high prices so as to take advantage of their technological advances and were subsequently compromised by an overall lack of results. The approach was revived when the tax code changed and the financial rewards of venture capitalists, coupled with the amazing string of technological breakthroughs achieved by small new companies, became apparent.[62]

For large firms, venture investment has allowed them to eat their cake and have it too. In the past, corporate giants tended to crush small, fragile new firms as they subsumed them, essentially killing the spirit and drive that had, in part, made them attractive acquisitions in the first place. Now, for an interest of 10 to 20 percent in the company, as well as aid in marketing and distribution (typically the most difficult challenges faced by start-ups), a large corporation such as General Electric or Raytheon—two of the biggest names in partnering—can own equity in potentially profitable companies. Moreover, such investors gain access to important new technology, not to mention the productive R & D capabilities for which new firms are now famous.[63]

In Washington County, this form of financing is not widespread but is practiced by at least one major firm, while the benefits of partnering have been felt by several smaller ones. In November 1983, Tektronix formed the Tektronix Development Corporation ostensibly for the purpose of making venture investments. In reality, it was forced into this move by what at the time was called "an embarrassing gush of venture-capital-fueled spin-offs."[64]

Tektronix Development's most important partnering venture by mid-1986 was TriQuint Semiconductor, a wholly owned subsidiary that began manufacturing high-speed integrated circuits on gallium-arsenide chips in 1985.[65] Founded by Alan D. Patz, a former employee, the new company leases space from Tektronix and is manufacturing its chips in the larger company's integrated circuit-fabrication plant at Beaverton.[66] While TriQuint has maintained an "arm's-length" relationship with Tektronix, its exploitation of gallium-arsenide technology, which was developed by its engineers when they were at Tektronix, has meant both prestige and potential profit for the parent firm.[67]

Tektronix has also made cash investments in Sequent Computer Systems and ATEQ Corporation.[68] Planar Systems, manufacturer of a flat-panel video display that uses Tektronix technology, has also received funds from the oscilloscope maker and from RCA.[69] Portland General Electric, a utility, has also opened a venture arm for partnering purposes.

By 1986 it was estimated that nearly $500 million in venture capital was available in the Pacific Northwest.[70] In great measure, the attraction to the region of this huge sum (which was not limited to electronics investments) was owing to the Washington County high-tech industry's maturation in the mid-seventies.[71] Between 1977 and 1986 at least seven venture companies had opened offices in Portland. These included the Shaw firm, underwritten by Portland's U.S. Bancorp First Source Capital, PacifiCorp Ventures, Northwest Venture Capital, Earl Kinship Capital, and Rainier Venture Partners. Other firms, such as Hambrecht and Quist, Crosspoint Venture Partners, and Sutter, Hill and Greylock Management regularly sent representatives to the Portland area and made significant investments in Washington County companies. At the same time, local stock brokerages and investment banking enterprises such as Paulson Investment Company entered the field.

In 1985, encouraged by the electronics industry's impact on the state's economy, the Oregon legislature created the Resource and Technology Development Corporation to nurture the industry. This gesture was originally opposed by the Oregon council of the American Electronics Association, which wanted government to stay out of the industry.[72] Preferring private investors with "deep pockets and long arms," as its executive director put it, the association nevertheless supported the corporation as the need became evident.[73]

The year 1983 was perhaps a watershed in venture funding for high technology, with venture moneys flowing in peak amounts to a variety of entrepreneurs.[74] However, financiers became more cautious as the market for initial public stock offerings became depressed in 1984 and persisted in remaining so.[75] As noted earlier, venture capitalists depend on successful stock offerings to generate the high-yielding, liquid returns that justify their initial investments.[76] The investing public's caution on new offerings has thus meant that much venture money is still locked up in older investments.[77] In turn, this has meant that mezzanine funding has been doled out more cautiously. However, some financiers have said that since 1983 the venture climate has been healthier because they have been forced to be more realistic about new entrepreneurial proposals.[78]

In a speech in 1981 to the Portland City Club, Ralph Shaw urged a policy of replacing the state's rapidly declining manufacturing base in forest products. He called for an aggressive program aimed at regenerating manufacturing employment by nurturing new firms, especially those in electronics.[79] Among many imaginative measures to achieve this end, he urged an expansion of the pool of venture funds.[80] It has been said that this speech was so bright, so obviously prescient that it changed the way the business community looked on new and small businesses and their relationship to the economy of the state.[81] What Shaw indicated about the combination of high technology and venture capital has come true in Washington County. For it is this linking of prudent investment with circuit boards in new and ever more creative schemes that has made a whole industry mushroom.

Epilogue

RARELY HAS THERE BEEN AN INDUSTRY as subject to rapid change as high technology. The postwar American automobile industry, for example, was for decades famous for transforming the looks of each new year's models, but such change was artificially induced and merely topical. The redesign of the cars' skins was actually a marketing strategy—planned obsolescence. Under the hood the products, the way they were made and their materials remained essentially the same year in and year out. Automakers never completely redesigned the car. Only the grim challenge of the Japanese imports in the eighties forced the U.S. industry out of its complacency. Then American cars were to some extent rethought and reconfigured in order to retain a share of their rapidly dwindling market.

The high-technology industry, on the other hand, has been forced constantly to respond and adapt to dramatic change. This is because the industry is intrinsically about change. The identity of high technology is, as noted at the beginning of this book, defined by the greater amount of research and development among its component firms compared with businesses in other sectors of the economy. Innovation is, therefore, a cornerstone of the industry. Moreover, the potential rewards for achieving innovation are so great and the costs associated with starting up in the field so negligible relative to other manufacturing businesses that young, energetic, and, most of all, imaginative engineers, scientists, and entrepreneurs have constantly entered the field to try to top a particular market. The frontiers of electronics have proved highly permeable. Advancements have come rapidly in the last thirty years, and one has seemingly led directly to another.

Innovation has, however, clearly been a double-edged sword for the industry. If each breakthrough has created a new market or revitalized an old one, it has also generated obsolescence for established products and, thus, fiscal instability for companies both young and old. The industry has also been more than usually susceptible to the vagaries of the world and national, not to mention regional, economies. The tendency of buyers of the industry's generally high-priced products to wait out recessions before purchasing new items, both out of concern for their own resources and in the hope and belief that new product lines would occur in the interim, has made recessions a dreaded event for electronics companies.

In any case, it can hardly be doubted that change—as both an economic and technological phenomenon—has periodically transformed high technology. By the end of the eighties, Washington County's high-technology community had felt both the lash and the balm of events generated by the fluctuations of the economy and by pathfinding research.

Among those companies suffering difficulties were Metheus, Lattice, Floating Point Systems, and Tektronix, to name only a few. Having survived the decade's first recession, they were nonetheless crippled by an industrywide downturn in 1985-86. Metheus, for instance, was just recovering from the slowdown by late 1987. It showed extraordinary increases in sales of computer-graphics display equipment, even reaching record levels.[1] The firm planned a stock offering in order to pay off pressing debts, using its glittering sales as an attraction to investors. But the stock market crash of 19 October 1987 made the offering untenable, and by spring 1988 the company was forced into Chapter 11 bankruptcy. This meant that it would stay in business while debt reorganization and a schedule of repayment to creditors was worked out by the Federal Bankruptcy Court. In order to reestablish its financial strength, the company planned to restructure its debt and focus on a new product, the Ultra Graphics Accelerator card for personal computers, which had had a highly favorable response from consumers.[2]

Floating Point Systems was also troubled late in the decade. A severe drop in business in the mid-eighties, coupled with the effects of the 1985-86 industry slowdown, left the company in a weakened condition. After Norman Winningstad quit FPS's day-to-day management in 1986, the team that replaced him was unable to sustain growth. Investor confidence in the company was diminished by misdevelopment, as well as a series of lawsuits filed in 1987 and 1988 charging that FPS officers had ignored or concealed the company's financial problems while making opportunistic business forecasts to the financial community, who then recommended FPS stock to investors. Disgruntled holders of FPS securities charged that they had been duped into paying "artificially inflated" prices for the stock of a company whose performance was at variance with descriptions.[3] The suits also alleged "massive insider trading" by a group of company officers.[4]

In mid-1988, while the suits were going through the courts, Floating Point Systems was forced to cut its work force by about one-third, leaving it with half the workers it had only a year earlier. However, some analysts thought that reorganization and fiscal retrenchment, begun in response to the problems facing the company, had created the conditions for full financial health and competitiveness by 1989.[5] The company had fallen behind in the minisupercomputer field, it was said, because its systems did not have the software variety available in those of the Convex and Alliant models.[6] Floating Point Systems' problems had been compounded when, by the end of 1986, only six of the T Series supercomputers had been delivered. Design and engineering inadequacies had given pause to both potential buyers and original purchasers. The T Series' flaws reflected the growing disarray of Floating Point Systems under its president, Lloyd Turner. Both Turner and the T Series were phased out, and Winningstad returned to head the company's day-to-day operations in October 1987.[7] His return, combined with a reshuffling of other major management responsibilities, was a sign that Floating Point's old aggressive and innovative style would be renewed.

Winningstad's departure from FPS had occurred in the first place because of the need for a firm hand at Lattice Semiconductor, a company in which he had made a sizable investment. Lattice had never recovered from the decade's economic disasters and the concomitant chip glut. Winningstad was confident that, through Chapter 11, the company's decline could be reversed, and by the time the filing was announced in July 1987, Lattice had experienced two quarters that showed slight profits.[8]

Some Washington County firms—mostly older, more established, and diversified—had experienced the same difficulties in the early years of the decade as the newer, more fragile companies, but had pulled directly out of them with the return of prosperity in the late eighties. ESI, for example, was forced in the last stages of high technology's recession of 1986-87 to cut some 30 percent of its work force to save money after posting the first loss in its history.[9] ESI was, however, well positioned to respond to an acute worldwide shortage of dynamic-random-access-memory chips, and this shored up its short-term business prospects enormously. Its customer base, which included IBM, Texas Instruments, Motorola, Samsung, and Fujitsu, were themselves well established, giving the company a reliable foundation on which to depend. Likewise, ESI's diversification worked well. The company made two laser machines used for chip manufacture for customers such as Ford, IBM, Digital, and Hughes Aircraft.

Similarly, Intel endured the high-technology slump to rebound in 1987 and 1988. With a reduced work force in the spring of 1987, the company was beginning to hire again and planned to open the two mothballed wafer plants in Aloha and new offices in Hillsboro.[10] "The last two years have been extremely difficult for our industry," said Robert Noyce in April; "however, we are at a point now where one can look at upgrading our facilities. Intel has been here [in Washington County] since 1976, and these investments are part of our long-term commitment to Oregon."[11]

At Tektronix the middecade slump hurt the company badly. Earl

Wantland retired in 1988 and the new management team undertook yet another reduction in force. The result was a leaner, stronger entity that the new leadership saw as being more able to cope with the competition.

Two Washington County companies that appeared to weather the economic storms of the eighties in stronger shape than before were the relative newcomers, Sequent and Mentor Graphics. In May 1987, Sequent was able to offer its stock to the public. Over-the-counter trading in Sequent securities subsequently raised some $29 million.[12] The public confidence that had brought this sum was based on its record of good management, as well as its leadership in supercomputers and parallel processing. On 18 May, the company introduced a new, higher speed computer that moved its products' computing speed past 80 million instructions per second.[13]

By 1987, Mentor had nearly doubled the market share of Daisy Systems in the computer-aided-engineering field.[14] Mentor's share was 32 percent, while its products were said to be used by twenty-three of the foremost twenty-five electronics companies.[15] Based on its great success in Asian and European markets, it had managed to continue its growth during the downturns while other American high-technology companies were losing money.[16] It planned to end the decade by diversifying its business into "computer-aided electronic packaging," the design of products like stereo systems (including the cabinetry), and computer-aided development of software.[17]

At the Oregon Graduate Center, meanwhile, research also continued apace. In 1988, two milestones were attained. Early in the year, the center announced that it had received a $5 million grant to start the Oregon Institute for Advanced Computing. The institute was to conduct research in areas such as artificial intelligence.[18] Later in the year, the center celebrated its twenty-fifth anniversary.

Such random sampling of the Washington County community gives a fair impression of the precarious, sometimes volatile, but always interesting, life of the high-technology industry. It is a story that vacillates between dramatic poles: Horatio Alger on one hand and Thomas Edison on the other. These are archetypal American economic and cultural myths, speaking to the traditional national journey toward material success and technological progress. The view that the person who advanced technology would become rich has historically been based on a solid core of truth. Edison himself is a classic example of its veracity. In both its tribulations and its triumphs Washington County's high technology resonates with the sounds that old dreams and new visions make. This is why it has figured so largely in the economy and society of the county and of Oregon.

A Note on the Sources

THIS NOTE EMPHASIZES our most important sources, but is not a comprehensive bibliography. For obvious reasons, business corporations do not make available to historians or other "outsiders" corporate records dealing with contemporary financial, patent, and technological data. However, officers and staff members of the high-technology firms we consulted did consent to be interviewed or to provide non-sensitive documents or both. Interviews were also conducted with many people in various fields outside business. Newspapers, particularly the *Oregon Journal*, the *Oregonian*, and the *Valley Times* were most helpful, especially in recording political developments. Business journals and magazines such as *Electronic Business*, *High Techonolgy*, the *Business Journal*, and *Oregon Business* were of great importance. So far as books are concerned, a picture of the West Coast high technology industry is drawn in Thomas Mahon, *Charged Bodies: People, Power and Paradox in Silicon Valley* (1985), and in Everett M. Rogers and Judith K. Larsen, *Silicon Valley Fever: Growth of High-Technology Culture* (1984). Marshall M. Lee, *Winning with People: The First 40 Years of Tektronix, 1946-1986* (1986), is indispensable.

Notes

BEGINNINGS

1. On the Tualatin, *see* Henry B. Zenk, "Contributions to Tualatin Ethnography: Subsistence and Ethno-Biology" (Master's thesis, Portland State University, 1976).
2. The coming of the Caucasians is described in Gordon B. Dodds, *The American Northwest: A History of Oregon and Washington* (Arlington Heights, Ill.: The Forum Press, 1986), 30-109, and Linda S. Dodds and Nancy A. Olson, *Cedar Mill History* (Portland, 1986).
3. For the influence of the gold rush, *see* Dodds, *The American Northwest*, 98-101.
4. Dodds, *The American Northwest*, 110-36; Dodds and Olson, *Cedar Mill History*.
5. Good surveys of the growth of the American high-technology industry are provided in Stan Augarten, *Bit by Bit: An Illustrated History of Computers* (New York: Ticknor and Fields, 1984); Thomas Mahon, *Charged Bodies: People, Power and Paradox in Silicon Valley* (New York: New American Library, 1985); Anthony Ralston and Edwin D. Reilly, Jr., eds., *Encyclopedia of Computer Science and Engineering*, 2nd ed. (New York: Van Nostrand Reinhold, 1983); Everett M. Rogers and Judith K. Larsen, *Silicon Valley Fever: Growth of High-Technology Culture* (New York: Basic Books, 1984).
6. Rogers and Larsen, *Silicon Valley Fever*, 37.

7. Augarten, *Bit by Bit*, 225.
8. Rogers and Larsen, *Silicon Valley Fever*, 38.
9. Mahon, *Charged Bodies*, 72.
10. The identity of the first automatic electronic digital computer is a matter of controversy. A case can be made that John V. Atanasoff, not John W. Mauchly, was the inventor. *See* Alice R. Burks and Arthur W. Burks, *The First Electronic Computer* (Ann Arbor: University of Michigan Press, 1988).
11. Robert Noyce, "Microelectronics," *Scientific American*, September 1977, 63-64.
12. Rogers and Larsen, *Silicon Valley Fever*, 105.

ELECTRO SCIENTFIC INDUSTRIES

1. Gary Craven Gray, *Radio for the Fireline: A History of Electronic Communication in the Forest Service, 1905-1975* (Washington, D.C.: U.S. Forest Service, 1982), 67-186.
2. Harold Lawson, interview with Linda S. Dodds, 10 February 1986.
3. Douglas Strain, interview with Linda S. Dodds, 29 January 1986, hereafter cited as Strain interview.
4. Bert Webber, *Retaliation: Japanese Attacks and Allied Countermeasures on the Pacific Coast in World War II* (Corvallis: Oregon State University Press, 1975).
5. Strain interview.
6. Strain interview; Broadcast Copyrights Incorporated, videotaped interview with Douglas Strain and other ESI employees, 1985, hereafter cited as BCI interview, ESI archives.
7. BCI interview.
8. BCI interview.
9. BCI interview.
10. ESI press release, 1966.
11. Strain interview; ESI, *Annual Report*, 1983, 10.
12. BCI interview.
13. Douglas C. Strain to Donald W. Pierce, 24 February 1966, ESI Archives; *Oregonian*, 5 August 1957; ESI, *Annual Report*, 1983, 10; ESI press release, 1966.
14. BCI interview.
15. BCI interview; Douglas McGregor, *The Human Side of Enterprise* (New York: McGraw-Hill, 1960).

16. Strain interviews, 9 and 29 January 1986; *Oregonian*, 9 September 1958; Strain to Pierce, 24 February 1966; *Oregon Journal*, 27 December 1959.
17. Strain interview, 29 January 1986; Rogers and Larsen, *Silicon Valley Fever*, 35-36; *Oregonian*, 21 December 1959.
18. Strain interview, 29 January 1986.
19. ESI press release, 1966; ESI, *Annual Report*, 1981.
20. Strain interview, 29 January 1986.
21. Strain to Pierce, 24 February 1966.
22. Strain interview, 29 January 1986; *Oregonian*, 19 July 1981.
23. Strain to Pierce, 24 February 1966; ESI, *Annual Report* 1983.
24. ESI, *Annual Report*, 1983.
25. ESI, *Annual Report*, 1983.
26. ESI, *Annual Report*, 1983.
27. "ESI/Palomar Acquires Capacitor Equipment," *Portland Magazine*, June 1984, 38; "ESI Acquires Palomar Systems and Machines Inc.," *Portland Magazine*, April 1982, 50; ESI, *Annual Report*, 1982.
28. Strain interview, 29 January 1986; *Wall Street Journal*, 26 August 1983; "ESI Files Public Offering Registration," *Oregon Business*, October 1983, 14; ESI, *Annual Report*, 1982.
29. BCI interview; ESI, *Annual Report*, 1983.
30. ESI, *Annual Report*, 1983.
31. Strain interview, 29 January 1986; Robert Levering, Milton Moskowitz, and Michael Katz, *The 100 Best Companies to Work For in America* (Reading, Mass.: Addison-Wesley, 1984), 102-104; *Oregonian*, 12 October 1985.

TEKTRONIX

This chapter was undertaken with the assistance of Professor Marshall M. Lee of Pacific University, archivist and historian for Tektronix. Although the corporate archives of Tektronix are not open to the public, where possible Professor Lee has provided references from the Tektronix archives.

1. Marshall M. Lee, *Winning With People: The First 40 Years at Tektronix, 1946-1986*, (Beaverton: Tektronix, 1986), 2-20.
2. Milt Bave, Logan Belleville, Miles Tippery, and Howard Vollum interviews with Marshall M. Lee, 24 June, 11 May, 25 June, 10 and 17 September 1984, respectively. These and all other interviews cited this chapter are in the Tektronix Archives, file 16.3:3.

3. Bave, Belleville, Tippery, and Vollum interviews.
4. Tekrad: Articles of Incorporation, 19 January 1946, Tektronix Archives, file 1.1:1.
5. Vollum interview.
6. Lee, *Winning With People*, 31.
7. Bave, Belleville, Tippery, and Vollum interviews.
8. Bill Hewlett and David Packard, Warren K. "Dal" Dallas interviews with Marshall M. Lee, 17 July, and 26 June 1984, respectively.
9. Lee, *Winning With People*, 31-33, 46-47.
10. Hewlett and Packard interview; Dallas interview.
11. Lee, *Winning With People*, 46.
12. Frank Hood and Dick Rhiger interview with Marshall M. Lee, 11 April and 16 May 1984, respectively.
13. Tektronix: Board of Directors Minutes, 10 January 1948. Tektronix Archives, file 2.1:1.
14. Bave, Belleville, Tippery, and Vollum interviews.
15. Tektronix: Board of Directors Minutes, 21 November 1947. Tektronix Archives, file 2.1:1.
16. Lee, *Winning With People*, 46.
17. Bave, Belleville, Tippery, and Vollum interviews.
18. Belleville and Rhiger interviews; Cliff Moulton, Bill Polits, and Dick Ropiequet interviews with Marshall M. Lee, 7 May, 9 May, and 24 March 1984, respectively.
19. Hood interview.
20. Derrol Pennington interview with Marshall M. Lee, 17 April 1984; Vollum, Hood, and Rhiger interviews.
21. Pennington, Vollum, Hood, and Rhiger interviews.
22. Lee, *Winning With People*, 51-52.
23. Pennington interview.
24. Lee, *Winning With People*, 51-52.
25. Lee, *Winning With People*, 261.
26. Lee, *Winning With People*, 48-49.
27. Tektronix: Board of Directors Minutes, 22 February 1949. Tektronix Archives, file 2.1:1.
28. Vollum interview.
29. Dallas and Vollum interviews; Bill Webber interview with Marshall M. Lee, 13 April 1984.
30. Lee, *Winning With People*, 49.
31. Lee, *Winning With People*, 83.
32. *Oregonian*, 21 June 1986.
33. Lee, *Winning With People*, 76.

34. Vollum interview.
35. Tektronix: Board of Directors Minutes, 26 April 1950. Tektronix Archives, file 2.1:1.
36. Lee, *Winning With People*, 82.
37. Webber and Pennington interviews; (Beaverton) *Valley News*, 9 August 1956.
38. Lee, *Winning With People*, 120-21.
39. Dallas, Vollum, Hewlett, and Packard interviews.
40. Dallas interview.
41. Dallas interview.
42. Lee, *Winning With People*, 98.
43. Hewlett and Packard interviews.
44. Lee, *Winning With People*, 160-62.
45. Lee, *Winning With People*, 161.
46. Dallas interview.
47. Dallas interview.
48. Lee, *Winning With People*, 194.
49. Lee, *Winning With People*, 172-78.
50. Lee, *Winning With People*, 159.
51. Lee, *Winning With People*, 194-97.
52. Lee, *Winning With People*, 198-202.
53. Lee, *Winning With People*, 245-48.
54. Vollum interview.
55. Lee, *Winning With People*, 248.
56. Vollum interview.
57. Lee, *Winning With People*, 221.
58. *Oregon Journal*, 17 May 1971; *Oregonian*, 18 May 1971.
59. *Wall Street Journal*, 11 January 1954.
60. Lee, *Winning With People*, 238.
61. Lee, *Winning With People*, 298.
62. Lee, *Winning With People*, 239-44.
63. Lee, *Winning With People*, 244-45.
64. Vollum interview.
65. Wim Velsink interview with Marshall M. Lee, 29 May 1984.
66. Lee, *Winning With People*, 257-66.
67. C. Norman Winningstad interview with Marshall M. Lee, 10 April 1984.
68. Lee, *Winning With People*, 278.
69. Lee, *Winning With People*, 282.
70. Lee, *Winning With People*, 283.
71. Lee, *Winning With People*, 284-85.

72. Lee, *Winning With People*, 291.
73. Paul Carlson, "In Memoriam," *Centerpost*, March 1986, 1.
74. *Oregonian*, 22 May 1986.
75. *Oregonian*, 22 May 1986.
76. *Oregonian*, 24 March 1985.
77. *Christian Science Monitor*, 12 May 1986.
78. *Oregonian*, 3 November 1985.

FLOATING POINT SYSTEMS

1. C. Norman Winningstad interview with Linda S. Dodds, 10 June 1986, hereafter cited as Winningstad interview.
2. Winningstad interview.
3. Winningstad interview.
4. Winningstad interview; Floating Point Systems, "Historically Speaking: A Chronological Account of Floating Point Systems, Inc. History, 1970-1983."
5. "Historically Speaking"; Winningstad interview.
6. Winningstad interview.
7. Winningstad interview.
8. Winningstad interview.
9. Winningstad interview; "Historically Speaking."
10. Floating Point Systems, *Annual Report*, 1978.
11. Floating Point Systems, *Annual Report*, 1978; Winningstad interview.
12. Floating Point Systems, *Annual Report*, 1979; "Historically Speaking"; *Oregonian*, 4 November 1980.
13. Floating Point Systems, *Annual Report*, 1980.
14. Floating Point Systems, *Annual Report*, 1980; Winningstad interview.
15. Floating Point Systems, *Annual Report*, 1981; Winningstad interview.
16. Winningstad interview.
17. Winningstad interview; Floating Point Systems, *Annual Report*, 1981.
18. *Oregonian*, 25 February 1984; *Forbes Magazine*, 18 June 1984; *Business Journal*, 28 January 1985; Susan E. Frost, "The Floating Point Clone," *Oregon Business*, January 1983, 78-83.
19. *Business Journal*, 11 March 1985; Floating Point Systems, *Annual Report*, 1985.
20. *Oregonian*, 27 July, 2 August 1986.

NEW VENTURES

1. *Business Journal*, 13 May 1986.
2. For background on the origin of Intel and the careers of its founders, *see* Tom Wolfe, "The Tinkerings of Robert Noyce," *Esquire*, December 1983, 346-74, and Glynnis Thompson Kaye, ed., *A Revolution in Progress... A History of Intel to Date* (Santa Clara: Intel, 1984).
3. Wolfe, "The Tinkerings of Robert Noyce," 354.
4. Kaye, *A Revolution in Progress...*, 11.
5. Kaye, *A Revolution in Progress...*, 11.
6. Kaye, *A Revolution in Progress...*, 8.
7. Wolfe, "The Tinkerings of Robert Noyce," 367.
8. Kaye, *A Revolution in Progress...*, 12.
9. Kaye, *A Revolution in Progress...*, 12.
10. Kaye, *A Revolution in Progress...*, 13.
11. Kaye, *A Revolution in Progress...*, 12.
12. Kaye, *A Revolution in Progress...*, 12.
13. George Drumbor interview with Craig Wollner, 6 November 1985, hereafter cited as Drumbor interview.
14. Kaye, *A Revolution in Progress...*, 9.
15. Drumbor interview.
16. Drumbor interview; *see also* Wolfe, "The Tinkerings of Robert Noyce," 368.
17. Drumbor interview; *see also Willamette Week*, 4 February 1980.
18. Drumbor interview; *Willamette Week*, 4 February 1980; *Oregonian*, 18 July 1978.
19. Drumbor interview.
20. Drumbor interview.
21. Drumbor interview.
22. *Oregonian*, 30 January 1974.
23. *Oregonian*, 30 January 1974.
24. *Oregonian*, 13 February 1977.
25. *Oregonian*, 13 February 1977.
26. *Oregonian*, 5 September 1979.
27. *Oregonian*, 5 September 1979.
28. *Oregonian*, 5 September 1979.
29. *Oregonian*, 18 July 1978.
30. *Oregonian*, 8 March 1982.
31. *Oregonian*, 8 March 1982.
32. Lee, *Winning With People*, 107-17.
33. Drumbor interview.

34. Drumbor interview.
35. *Oregonian*, 5 November 1985.
36. *Oregonian*, 5 November 1985.
37. *Oregonian*, 5 November 1985.
38. *Oregonian*, 5 November 1985.
39. *Oregonian*, 11 October 1985.
40. *Oregonian*, 5 November 1985.
41. *Oregonian*, 17 April 1980.
42. *Oregonian*, 2 September 1982, 13 December 1985, 12 April 1986, 27 July 1986.
43. *Oregonian*, 12 April 1986.
44. Ron Karten, "Preparing for a Big Takeoff," *Oregon Business*, September 1982, 66-72.
45. Karten, "Preparing for a Big Takeoff," 66-72.
46. Karten, "Preparing for a Big Takeoff," 66-72.
47. "Flight Dynamics Undergoes Dramatic Changes in 1982," *Oregon Business*, February 1983, 12-13.
48. "Flight Dynamics Undergoes Dramatic Changes in 1982," 12-13.
49. Flight Dynamics, *Annual Report*, 1985.
50. Flight Dynamics, press releases, 2 May, 3 June 1986.
51. Sharon VanSickle and Edwin Lee, "Mentor Graphics Corporation, Press Background Information," 15 October 1985.
52. Gary Eisler, "From Foxholes to Ferraris," *Oregon Business*, June 1985, 23.
53. Tom Bruggere interview with Craig Wollner, 13 May 1986, hereafter cited as Bruggere interview; Eisler, "From Foxholes to Ferraris," 22.
54. Bruggere interview.
55. Mentor Graphics, *Annual Report*, 1984.
56. Mentor Graphics, *Annual Report*, 1984.
57. VanSickle and Lee, 14.
58. VanSickle and Lee, 14.
59. VanSickle and Lee, 15.
60. VanSickle and Lee, 15-18.
61. VanSickle and Lee, 1.
62. Bruggere interview.
63. VanSickle and Lee, 6-12.
64. VanSickle and Lee, 12-13.
65. VanSickle and Lee, 12.
66. Bruggere interview.
67. Bruggere interview.
68. Bruggere interview.

69. Gail Kinsey Hill, "Computer-Aided Everything," *Oregon Business*, April 1985, 16.
70. Hill, "Computer-Aided Everything," 16.
71. Lewis N. Goslin interview with Craig Wollner, 29 May 1986, hereafter cited as Goslin interview.
72. Cindy Thames, "Engineering Workstations: The Dash to the Desktop," *Electronic Business*, September 1983, 92.
73. Goslin interview.
74. Bruggere interview; *see also* Thames, "Engineering Workstations: The Dash to the Desktop," 91-92.
75. Goslin interview.
76. Alden M. Hayashi, "Mentor at a Crossroad: Picking Another Platform," *Electronic Business*, 15 January 1986, 42-43.
77. Bruggere interview; Goslin interview.
78. Goslin interview.
79. Goslin interview.
80. Goslin interview.
81. *See* Hill, "Computer-Aided Everything," 15-18.
82. Bruggere interview.
83. Bruggere interview.
84. Bruggere interview.
85. Bruggere interview.
86. Metheus Corporation, "Background Information on Metheus Corporation."
87. Metheus Corporation, "Background Information on Metheus Corporation."
88. Metheus Corporation, "Background Information on Metheus Corporation."
89. "CAE Battleground Shapes Up in Oregon," *Oregon Business*, July 1984, 12-16; *Oregonian*, 5 June 1984; Norm Alster, "David and Goliath Join Hands in Design Market," *Electronic Business*, 1 September 1984, 168-70.
90. *Electronic Business*, 1 April 1986, 74-75.
91. Metheus Corporation, "Background Information on Metheus Corporation."
92. James W. Hurd interview with Gordon B. Dodds, 17 December 1985, hereafter cited as Hurd interview.
93. Hurd interview; *Electronic Business*, May 1981, 74; Frank Catalano, "Display Diagnosis: Good News for a Flat Business," *Electronic Business*, November 1983, 222-24; James W. Hurd to Gordon B. Dodds, 12 October 1987.

94. *Electronic Business*, May 1981, 74; Catalano, "Display Diagnosis," 222-24; *High Technology*, December 1985, 42; Hurd interview.
95. Hurd interview; Hurd to Dodds, 12 October 1987.
96. *Electronic Business*, May 1981, 74; Catalano, "Display Diagnosis," 222-24.
97. *High Technology*, December 1985, 42-45.
98. (Tigard) *Valley Times*, 19 December 1985.
99. Hurd to Dodds, 12 October 1987.
100. *Oregonian*, 7 May 1986.
101. *Oregonian*, 26 January 1986.
102. *Oregonian*, 11 May 1985, 27 July 1986; Winningstad interview.
103. *Oregonian*, 26 January, 27 July 1986; Ed Coughlin to Gordon B. Dodds, 22 August 1989.
104. Tim Mead, "Patience Pays Off For NEC America," *Electronic Business*, 1 February 1985, 56-62.
105. Mead, "Patience Pays Off For NEC America," 56-62; *Oregonian*, 29 September 1984.
106. *Oregonian*, 22 May 1984; Jessica Schwartz, "The Three-pronged Strategy of Epson's Yasuhiro Tsubota," *Electronic Business*, 1 March 1985, 128-33; NEC, "This Is NEC 1985."
107. Pat Christiansen to Gordon B. Dodds, 10 July 1986; *Oregonian*, 9 March 1986; 21 May 1986.
108. *Oregonian*, 28 August, 11 October 1984; Schwartz, "The Three-pronged Strategy of Epson's Yasuhiro Tsubota," 128-33.
109. Epson Portland, *In Progress*, November 1985.
110. Epson Portland, press release, July 1985.
111. *Oregonian*, 12 September 1984, 15 January 1985, 8 April 1986; Mike Brennan, "Fujitsu's Goal This Year: 50% Growth in the U.S. Market," *Electronic Business*, 1 August 1984, 46-48.
112. Epson Portland, press release, July 1985.
113. *See* chapter 8.
114. Governor's Office, press release, 6 March 1986.
115. Economic Development Department, "Cancun Background Summary," 21 May 1984.
116. David Lutjen, "Opening an Office to Foreign Doors," *Oregon Business*, March 1985, 39-45; *Oregonian*, 9 July 1986.
117. Robert L. Hill, "Japan's Yen For Oregon," *Oregon Business*, March 1985, 33-37; Governor's Office, press release, 6 March 1986; Gov. Victor Atiyeh, speech at Portland State University, 14 June 1984.

THE SUPERCOMPUTERS

1. Dan Minte, "Supercomputers," *Oregonian, Northwest Magazine*, 11 May 1986, 6-10, 22.
2. Winningstad interview; Cornell University, Center for Theory and Simulation in Science and Engineering, *Forefronts*, October and November 1985, May 1986.
3. *Oregonian*, 3 April 1986; (Tigard) *Valley Times*, 10 April 1986.
4. Minte, 6-10; Floating Point, press releases, 7 April, 14 May 1986.
5. Peter Denning, "The Science of Computing: The Evolution of Parallel Processing," *American Scientist*, August 1985, 322-23; Intel, *iPSC System Product Summary* (n.d., n.p.).
6. *NCube Ten; Electronics*, 18 November 1985, 23-24; *VLSI Systems Design*, March 1986, 18, 111; Terry Costlow, "64 uPs, hypercube design control agile robot," *Electronic Design*, 3 April 1986, 33-34.
7. *Oregonian*, 13 January 1985; Scientific Computer Systems, "Backgrounder," 10 March 1986; Stephen Ohr, "Computer Scores with Cray Compatibility," *Electronic Design*, 6 March 1986; Minte, 6-10; Nancy Hays, "New Systems Offer Near Supercomputer Performance," *Computer*, March 1986, 104-107; *Oregonian*, 12 August 1986.
8. Gregory Johnsen, "Gallium Arsenide Chips Emerge From the Lab," *High Technology*, July 1984, 44-51.
9. *Electronic Business*, October 1982, 44; David Card, "GaAs Could Be A Boon to Microwave Instruments," *Electronic Business*, 1 May 1986, 30.
10. *VLSI Design*, July 1985.
11. *VLSI Design*, July 1985 (quotation); *Defense Electronics*, April 1986, 45.
12. Theresa Engstrom, "Marriage, Silicon Style," *CEO*, May 1985, 33-38; Bipolar Integrated Technology, "Bipolar Integrated Technology Backgrounder."
13. *Electronic Engineering Times*, 14 April 1986, 30; Bipolar Integrated Technology news release, 10 April 1986; *Computer Design*, 15 April 1986.
14. Casey Powell interview with Gordon B. Dodds, 19 December 1986; *Oregonian*, 9 September 1984.
15. *Oregonian*, 18 March 1983.
16. *Oregonian*, 9 September 1984.
17. Sequent press release, 8 July 1985.
18. *Oregonian*, 2 March 1986.
19. *Oregonian*, 26 December 1985.

HIGH TECHNOLOGY AND EDUCATION

1. Mark O. Hatfield, *Not Quite So Simple* (New York: Harper and Row, 1968), 52-63; State of Oregon, Governor's Committee on Science, Engineering, and New Technologies, "Report on Portland Center for Graduate Study," 26 May 1961.
2. "Concepts and Organization of Graduate Center, An Advisory Committee Report to the Governor of Oregon," 30 July 1962.
3. "Concepts and Organization of Graduate Center, An Advisory Committee Report to the Governor of Oregon," 30 July 1962.
4. *Oregon Laws*, 1963, 1376.
5. Portland City Club, "Report on Graduate Education and Research Facilities for Metropolitan Portland," *Portland City Club Bulletin*, 10 May 1963.
6. *Portland City Club Bulletin*, 10 May 1963.
7. James Hurst interview with Gordon B. Dodds, 6 August 1986, hereafter cited as Hurst interview; Steve Dodge, "25 Years of Science Inquiry at OGC," part 1, *Visions*, (Spring 1988), 16.
8. Dodge, "25 Years of Science Inquiry at OGC," part 1, 16; Frederick Holmes interview with Gordon B. Dodds, 29 July 1986, hereafter cited as Holmes interview.
9. Holmes interview.
10. Bert Gredvig interview with Linda S. Dodds, 13 May 1986, hereafter cited as Gredvig interview; Oregon Graduate Center, 1969-1970 Catalog.
11. Holmes interview; Hurst interview.
12. Richard Elliott interview with Gordon B. Dodds, 4 August 1986, hereafter cited as Elliott interview; Dodge, "25 Years of Science Inquiry at OGC," part 1, 17.
13. Gredvig interview; Hurst interview; Dodge, "25 years of Science Inquiry at OGC," part 2, *Visions*, (Summer 1988), 19-20.
14. Elliott interview; Hurst interview.
15. *Oregonian*, 30 April 1978, 2 May 1978; Maryanne Crane interview with Gordon B. Dodds, 6 August 1986, hereafter cited as Crane interview.
16. Elliott interview; James Huntzicker interview with Gordon B. Dodds, 7 August 1986, hereafter cited as Huntzicker interview; Crane interview; *Business Week*, 14 April 1975, 70ff.
17. Hurst interview.
18. *Valley Times*, 20 September, 9 December 1973; 10 January, 17, 28 February 1974; *Oregon Journal*, 17 December 1974.

19. Crane interview; Elliott interview; *Oregonian*, 11, 13 June 1978; *Oregon Journal*, 6 July 1978.
20. F. Paul Carlson interview with Linda S. Dodds, 11 December 1985, hereafter cited as Carlson interview.
21. Carlson interview; Holmes interview.
22. Carlson interview; *Oregonian*, 21 May 1982; *Oregon Journal*, 21 May 1982; *Valley Times*, 26 May 1982.
23. Carlson interview; Oregon Graduate Center Catalogs, 1980, 1986.
24. Carlson interview; *Valley Times*, 29 August 1985; *Oregonian*, 7 May 1986.
25. Huntzicker interview; Oregon Graduate Center Catalogs, 1980, 1986.
26. Oregon Graduate Center Catalog, 1985-86.
27. *Visions*, Summer 1985; Spring 1986; Fall 1986.
28. Pat McCormick interview with Gordon B. Dodds and Craig E. Wollner, 22 July 1986, hereafter cited as McCormick interview; Oregon Council, American Electronics Association, "A Review of the Oregon Council's Experiences Managing Government Affairs, 1983 to Present;" Roger Olsen interview with Gordon B. Dodds, 4 February 1986, hereafter cited as Olsen interview; Barbara Edge, "High Tech's Education Ante," *Oregon Business*, March 1983, 66-69.
29. Edge, "High Tech's Education Ante," 66-69.
30. Olsen interview.
31. Oregon Educational Coordinating Commission High Technology Grant Program, "Project Summary."
32. William E. Davis, "Proposal for a High Technology Engineering/Computer Science Educational Program for Industry."
33. Davis, "Proposal for a High Technology... ," 5-6.
34. Davis, "Proposal for a High Technology... ," 9, 11, 23-63.
35. Remarks by Governor James B. Hunt, Jr., 17 August 1985.
36. "Legislators Listen to Education Message," *Oregon Business*, May 1983, 21.
37. Gail Kinsey Hill, "Washington's High Tech Center: Lure or a Lemon," *Oregon Business*, April 1984, 73-74; Gov. Victor Atiyeh, "State of the State" speech, 7 March 1985.
38. McCormick interview; Oregon Council, American Electronics Association *Review*.
39. Governor's Executive Order No.85-9; Governor's Office, press release, 11 September 1985; Oregon Center for Advanced Technology Education, "A Statement of Mission and Strategic Objectives," 9 January 1986, quotations, 2-3; "Operational and Implementation Plan."
40. Olsen interview; Oregon Center for Advanced Technology Education,

"OCATE"; Oregon State System of Higher Education, Board Minutes, 26 April, 10 September 1985, 21 February 1986.

41. Gary Conkling interview with Craig Wollner, Spring 1986, hereafter cited as Conkling interview.
42. Conkling interview; Lintner Center, "Lintner Center"; Lintner Center, "The Paul D. Lintner Center for Advanced Education."
43. Conkling interview; *Valley Times*, 5 December 1985; 23 January and 12 June 1986.
44. ESI, various newsletters; *Texas Aggie*, October 1983, 34-35; *Solid State Technology*, January 1984, 40; Floating Point Systems, *To The Point*, 28 March 1980; "UO and OSU Institute Moves Toward Fruition," *Oregon Business*, June 1984, 14-16; "Oregon Engineers Show Job Loyalty," *Oregon Business*, May 1985, 15, 102.
45. "New Graduate Programs Offered," *Oregon Business*, September 1985, 14, 86; "PCC Offers Electronics Degree," *Oregon Business*, August 1982, 15.

POLITICS

1. McCormick interview; *Oregonian*, 9 June 1981.
2. Rogers and Larsen, *Silicon Valley Fever*, 63-64; Mahon, *Charged Bodies*, 97.
3. Rogers and Larsen, *Silicon Valley Fever*, 63-64.
4. Rogers and Larsen, *Silicon Valley Fever*, 65.
5. 76 U.S. Stat., 217.
6. *Oregonian*, 22 January 1986.
7. Mike Malone, "New Facts in the Fractious Flap Over the Flat Tax," *Electronic Business*, 15 February 1985, 67-68.
8. Malone, "New Facts in the Fractious Flap Over the Flat Tax," 67-68; *Oregonian*, 9 January 1986.
9. *Christian Science Monitor*, 25 February 1986; *Oregonian*, 22 January 1986.
10. *Valley Times*, 6 February 1986; 31 July 1986.
11. *Electronic Business*, 1 May 1983, 78; Nancy Henderson, "GSP, a Duty-Free-Import Program, Draws EIA Fire," *Electronic Business*, July 1983, 58.
12. Jonathan B. Tucker, "High Tech Firms Push for New Export Control Rules," *High Technology*, December 1984, 83, 86.
13. "Operation Exodus Causing Problems for Oregon Firms," *Oregon Business*, March 1983, 12, 14.

14. "Operation Exodus Causing Problems for Oregon Firms," 14; Jon Zonderman, "New Law May Spur Technology Exports," *High Technology*, April 1983, 84-85.
15. 98 U.S. Stat., 1815; Mary Jo Foley and Nancy Henderson, "Cooperative Research: We Won, But Who Cares?" *Electronic Business*, 1 February 1985, 39-40.
16. Shelley Tsantes, "High Tech Confronts a Hazardous Dilemma, *Electronic Business*, 1 March 1986, 36-44.
17. Tsantes, "High Tech Confronts a Hazardous Dilemma," 56.
18. Daniel Gottlieb, "High Tech on Capitol Hill," *Electronic Business*, 15 August 1985, 122.
19. *Electronic Business*, June 1981, 46, 49.
20. *Oregonian*, 27 June 1986.
21. Nancy Ryles interview with Linda S. and Gordon B. Dodds, 23 January 1986, hereafter cited as Ryles interview.
22. *Oregonian*, 31 July 1983; McCormick interview; Oregon Council of the American Electronics Association, "A Review of the Oregon Council's Experiences Managing State Government Affairs 1983 to Present."
23. Dodds, *The American Northwest*, 324.
24. *Oregonian*, 31 July 1983.
25. 1983 *Oregon Laws* Ch. 814; SB 568; SB 245; SB 294.
26. American Electronics Association "Review."
27. American Electronics Association "Review."
28. Bill Mainwaring, "The Unitary Tax: Method or Madness," *Oregon Business*, April 1984, 26.
29. Thames, "Engineering Workstations: The Dash to the Desktop," 54-56; Dennis Livingston, "States Urged to Soften Unitary Tax Bite," *High Technology*, October 1984, 98-99.
30. McCormick interview; Mainwaring, "The Unitary Tax: Method or Madness," 26-27.
31. "Speech by Governor Victor Atiyeh [to] the World Affairs Council, Portland, June 20, 1984."
32. *Oregonian*, 21 June 1984.
33. (Medford) *Mail-Tribune*, 28 June 1984.
34. *Oregonian*, 6 July 1984; (Tigard) *Valley Times*, 26 July 1984.
35. (Klamath Falls) *Herald and News*, 31 July 1984.
36. Ryles interview.
37. *Wall Street Journal*, 1 August 1984; (Salem) *Statesman-Journal*, 23 August 1984; *Oregonian*, 5 August 1984; *Herald and News*, 3 August 1984.

38. American Electronics Association "Review"; McCormick interview.
39. McCormick interview; American Electronics Association "Review"; "New Graduate Program Offered," *Oregon Business*, September 1985, 14.
40. American Electronics Association "Review."
41. American Electronics Association "Review."
42. American Electronics Association "Review."
43. McCormick interview; American Electronics Association "Review"; 1985 *Oregon Laws*, ch. 726; Richard P. Reiter interview with Gordon B. Dodds, 28 January 1986, hereafter cited as Reiter interview.
44. American Electronics Association "Review"; 1985 *Oregon Laws*, chs. 735, 737.
45. American Electronics Association "Review"; McCormick interview.
46. *Oregon Journal*, 14 August 1959, 12 December 1963; Tom McCall, *Tom McCall: Maverick, An Autobiography with Steve Neal* (Portland: Binford and Mort, 1977), 190-191; Charles Deemer, "A Conversation With Tom McCall," *Oregon Business*, September 1981, 16-26.
47. Robert L. Hill, "Leaderless Crusade for Economic Development," *Oregon Business*, September 1983, 26-30.
48. 1983 *Oregon Laws*, ch. 317.
49. Robert L. Hill, "On the Brink," *Oregon Business*, February 1984, 22-24, 65-67.
50. *High Technology*, January 1985, 27.
51. *High Technology*, January 1985, 27, 28.
52. 1985 *Oregon Laws*, ch. 777.
53. 1985 *Oregon Laws*, ch. 777.
54. Wes Myllenbeck, interview with Gordon B. and Linda S. Dodds, 6 February 1986, hereafter cited as Myllenbeck interview.
55. Myllenbeck interview.
56. Myllenbeck and Strain interviews.
57. Ryles interview.
58. Hurd interview.
59. Unified Sewerage Agency, *Annual Report*, 1985; *Strides*, second quarter, 1976, 1-3.
60. *Valley Times*, 24 July, 31 July 1986; *Oregonian*, 14 June 1986.
61. Myllenbeck interview.
62. Metropolitan Service District, "Metro's Urban Growth Boundary: Background and Explanation," (n.d., n.p.); *Oregonian*, 13 May 1986.
63. Gary Conkling interview with Gordon B. Dodds, 7 July 1986, hereafter cited as Conkling interview.; *Valley Times*, 7 August 1986.

64. *Girt et al.* v. *Tri-Met et al.*, 4 Oregon Tax Report 92.
65. Conkling interview.
66. Elizabeth J. Radigan interview with Gordon B. Dodds, 12 December 1985.

WORLD OF WORK

1. *See* Richard W. Riche, Daniel E. Hecker, and John V. Brugan, "High Technology Today and Tomorrow: A Small Slice of the Employment Pie," *Monthly Labor Review*, November 1983, 54.
2. Riche, et al., "High Technology Today and Tomorrow: A Small Slice of the Employment Pie," 51-53; John V. Brugan, "Cyclical Behaviour of High Tech Industries," *Monthly Labor Review*, May 1985, 9.
3. Riche, et al., "High Technology Today and Tomorrow," 58.
4. Riche, et al., "High Technology Today and Tomorrow," 58.
5. Bruggere interview; *see also* Ralph Shaw interview with Craig Wollner, hereafter cited as Shaw interview, 22 May 1986, for a slightly different interpretation.
6. *Oregonian*, 21 May 1986.
7. *Oregonian*, 21 June 1986.
8. *Oregonian*, 29 December 1985.
9. John Stowe and Alan Stoebig, "Portland Metropolitan Labor Trends," State of Oregon Employment Division, April 1986, 4.
10. Stowe and Stoebig, "Portland Metropolitan Labor Trends," 4.
11. Stowe and Stoebig, "Portland Metropolitan Labor Trends," 4.
12. Stowe and Stoebig, "Portland Metropolitan Labor Trends," 4.
13. Stowe and Stoebig, "Portland Metropolitan Labor Trends," 4.
14. Stowe and Stoebig, "Portland Metropolitan Labor Trends," 4.
15. *Valley Times*, 30 January 1986.
16. *Valley Times*, 30 January 1986.
17. *Valley Times*, 6 February 1986.
18. *Valley Times*, 6 February 1986.
19. Gail Kinsey Hill, "Changes on the Line," *Oregon Business*, May 1985, 24.
20. Hill, "Changes on the Line," 24.
21. Hill, "Changes on the Line," 24.
22. *Valley Times*, 2 February 1986.
23. *Valley Times*, 2 February 1986.
24. Drumbor interview.

25. E.g., Ralph Shaw, "Capital Formation and Economic Development in Oregon," Speech to the City Club of Portland, 9 October 1981.
26. John Hoer, "Beyond Union," *Business Week*, 8 July 1985, 74.
27. Theresa Engstrom "Union Dues: High Tech Meets Organized Labor," *Electronic Business*, December 1983, 100.
28. Engstrom, "Union Dues," 100.
29. *See* William E. Leuchtenberg, *A Troubled Feast: American Society Since 1945* (Boston: Little, Brown, 1973), 6.
30. Wolfe, "The Tinkerings of Robert Noyce," 346-74.
31. Lee, *Winning With People*, 38.
32. Lee, *Winning With People*, 38, 49.
33. Lee, *Winning With People*, 39.
34. Lee, *Winning With People*, 49.
35. Lee, *Winning With People*, 49.
36. William B. Glaberson, "Rolling Back the Boss's Right to Fire at Will," *Business Week*, 8 July 1985, 74-75.
37. Ralph Wiser interview with Craig Wollner, hereafter cited as Wiser interview, 29 July, 1986.
38. Wiser interview.
39. Glaberson, "Rolling Back the Boss's Right to Fire at Will," 74.
40. Linda Rasmussen interview with Craig Wollner, 14 November 1985, hereafter cited as Rasmussen interview. *See also* Engstrom, "Union Dues," 102-103.
41. Rasmussen interview.
42. Rasmussen interview.
43. Rasmussen interview.
44. Rasmussen interview.
45. Rasmussen interview.
46. *Willamette Week*, 17 October 1985.
47. *Willamette Week*, 17 October 1985.
48. *Willamette Week*, 17 October 1985.
49. Rasmussen interview.
50. Rasmussen interview.
51. Rasmussen interview.
52. Engstrom, "Union Dues," 102.
53. Engstrom, "Union Dues," 102.
54. Engstrom, "Union Dues," 102.
55. Engstrom, "Union Dues," 102.
56. "Intel Has a Problem Hiring Engineers," *Oregon Business*, August 1984, 43.

57. "Intel Has a Problem Hiring Engineers," 43.
58. "Intel Has a Problem Hiring Engineers," 43.
59. *Occupational Employment Trends*, District 2, (Oregon State Employment Division, 1984), 53.
60. Bruggere interview.
61. "Electronics Employment to Grow 43 Percent," *Oregon Business*, May 1983, 13-14.
62. "Oregon Engineers Show Job Loyalty," *Oregon Business*, May 1985, 104.
63. Wolfe, "The Tinkerings of Robert Noyce," 346-374.
64. Wolfe, "The Tinkerings of Robert Noyce," 346-374.
65. Wolfe, "The Tinkerings of Robert Noyce,"346-374.
66. Robert Levering, Milton Moskowitz, and Michael Katz, *The 100 Best Companies To Work For In America* (Reading, Mass.: Addison-Wesley, 1984), 152.
67. Levering et al., *The 100 Best Companies*, 154.
68. Bruggere and Drumbor interviews.
69. Bruggere interview.
70. Bruggere interview.
71. Shaw interview.
72. "Intel Has a Problem Hiring Engineers," 43.

THE BALANCE SHEET

1. *Valley Times*, 6 October 1985.
2. *Valley Times*, 6 October 1985.
3. Ron Karten, "Industrial Parks: Changing With the Times," *Oregon Business*, March 1984, 41.
4. Karten, "Industrial Parks," 41. *See also Valley Times*, 12 December 1985.
5. Karten, "Industrial Parks," 41.
6. Karten, "Industrial Parks," 41.
7. Karten, "Industrial Parks," 41.
8. Karten, "Industrial Parks," 42.
9. Karten, "Industrial Parks," 42.
10. Ron Karten, "A Cloudy Sunset?" *Oregon Business*, April 1986, 52.
11. Karten, "Cloudy Sunset," 52.
12. Karten, "Industrial Parks," 42.
13. Karten, "Industrial Parks," 45.
14. Karten, "Industrial Parks," 45.

15. Karten, "Industrial Parks," 45.
16. Karten, "Industrial Parks," 42.
17. Karten, "Industrial Parks," 42.
18. Karten, "Industrial Parks," 45.
19. Leland and Hobson, [economics consulting] "The Need For Additional Land in the Sunset Corridor," October 1985, 10.
20. Leland and Hobson, "The Need For Additional Land" 10.
21. *Beaverton Business Advocate*, May 1986, 3.
22. M.J. Murdock Charitable Trust, *Annual Reports*, 1975-1976; 1984.
23. *Oregonian*, 25 April 1986; Dolores Winningstad interview with Linda S. Dodds, 17 January 1986; *Oregonian*, April 1986, 58.
24. *Valley Times*, 6 February 1986; *Tekweek*, 5 February 1986.
25. Rogers and Larsen, *Silicon Valley Fever*, 197-99; Diana Hembree, "Dead End in Silicon Valley: The High Tech Future Doesn't Work," *Progressive*, October 1985, 18-24; Nancy Henderson, "Welcome to Silicon Valley But Please Don't Drink the Water," *Electronic Business*, June 1983, 78-79.
26. *Valley Times*, 15, 29 August 1985; Reiter interview.
27. Reiter interview.
28. Reiter interview.
29. *Valley Times*, 15 August 1986; "Dr. James Pankow: Protecting Our Groundwater," *Visions*, Summer 1985, 10-15.
30. Ryles interview; Mary Jo Foley, "Women: Most Underrepresented Minority in Electronics," *Electronic Business*, 15 October 1984, 120, 122. *Annual Reports 1985* of ESI, Flight Dynamics, Floating Point Systems, Mentor Graphics, Oregon Graduate Center, and Tektronix.
31. Foley, "Women: Most Underrepresented Minority in Electronics," 108-109, 120, 122.
32. Steve Karakas interview with Gordon B. Dodds, 18 March 1986; *Business Journal*, 16 July 1984; Karakas, Van Sickle, Ouellette, "Client List, February 14, 1986."
33. Jim Buchanan interview with Gordon B. Dodds, 11 March 1986; The Waggener Group, "Client Contacts"; *Business Journal*, 16 July 1984.
34. *Oregonian*, 15 December 1985.
35. G. Pascal Zachary, "High Tech's New Reality," *Willamette Week*, 17 October 1985; Perkins, Coie, "Perkins, Coie" (booklet).
36. Herbert Swartz, "Marriage of Convenience: Computer Vendors and Dealers," *Electronic Business*, October 1983, 98; Robert L. Birnbaum, "Keeping the Secrets All in the Family," *Electronic Business*, 15 February 1985, 22, 24; *High Technology*, February 1986, 6; Perkins, Coie, "Perkins, Coie."

37. Chris Sturm, "Portland Emerging as a Venture Capital Center," *Portland Magazine*, September 1984, 45.
38. Ralph Shaw, "Capital Formation and Economic Development in Oregon," Speech to the Portland City Club, 9 October 1981.
39. David Olson, "Venture Capital: Financing High Tech," *Oregon Business*, July 1982, 46.
40. *Oregonian*, 11 May 1986; *Business Journal*, 16 September 1985.
41. Sturm, "Venture Capital," 45.
42. Robert Heilbroner, *The Making of Economic Society* (Englewood Cliffs, N.J.: Prentice Hall, 1962), 151.
43. Sturm, "Venture Capital," 45.
44. Wolfe, "The Tinkerings of Robert Noyce," 346-374; Bruggere interview.
45. *Oregonian*, 11 May 1986.
46. Michael Malone, "Successful High Tech Start-Ups Require More Than Innovation," *Electronic Business*, 1 April 1981, 100.
47. Sturm, "Venture Capital," 45.
48. Shaw interview.
49. Shaw interview.
50. Shaw interview.
51. Shaw interview.
52. Shaw interview.
53. *Business Journal*, 3 March 1986.
54. *Business Journal*, 3 March 1986.
55. *Business Journal*, 16 September 1985.
56. *Business Journal*, 16 September 1985.
57. *Business Journal*, 16 September 1985.
58. *Business Journal*, 16 September 1985.
59. *Business Journal*, 16 September 1985.
60. *Business Journal*, 16 September 1985.
61. *Christian Science Monitor*, 15 April 1986, 21-22.
62. *Christian Science Monitor*, 15 April 1986, 21-22.
63. *Christian Science Monitor*, 15 April 1986, 21-22.
64. *Oregonian*, 24 March 1985.
65. *Oregonian*, 24 March 1985.
66. *Oregonian*, 24 March 1985.
67. *Oregonian*, 24 March 1985.
68. *Oregonian*, 24 March 1985.
69. *Oregonian*, 24 March 1985.
70. Shaw interview.
71. *Business Journal*, 16 September 1985.

72. McCormick interview.
73. McCormick interview.
74. *Business Journal*, 16 September 1985.
75. *Business Journal*, 16 September 1985.
76. *Business Journal*, 16 September 1985.
77. *Business Journal*, 16 September 1985.
78. *Business Journal*, 16 September 1985.
79. Shaw, "Capital Formation," speech to the Portland City Club.
80. Shaw, "Capital Formation," speech to the Portland City Club.
81. William Glasgow interview with Craig Wollner, 5 February 1986.

EPILOGUE

1. *Oregonian*, 21 September 1987.
2. *Oregonian*, 23 April 1988.
3. *Oregonian*, 25 March 1988.
4. *Oregonian*, 25 March 1988.
5. *Oregonian*, 7 June 1987.
6. *Oregonian*, 7 June 1987.
7. *Willamette Week*, 11 December 1986.
8. *Oregonian*, 3 July 1987.
9. *Oregonian*, 10 April 1988.
10. *Oregonian*, 29 April 1987.
11. *Oregonian*, 29 April 1987.
12. *Oregonian*, 1 May 1987.
13. *Oregonian*, 1 May 1987.
14. *Oregonian*, 30 January 1987.
15. *Oregonian*, 30 January 1987.
16. *Oregonian*, 30 January 1987.
17. *Oregonian*, 30 January 1987.
18. *Oregonian*, 9 January 1988.

Index

AFL-CIO, 138
AMOCO, 49
ARNAV Systems, 66
ATEQ Corporation, 179, 186
Advisory Committee on the Commission on Technical Education, 119
Aerojet General, 75
Aikens, Howard, 8
Airborne Express, 65
Aloha, Oregon, 61, 62, 191
Amdahl, 87
American Electronic Association, 56, 57, 125, 134, 141-45 *passim*, 160, 163, 166:
  Oregon Council, 112, 113, 116, 118, 134, 140, 146, 186
American Telephone and Telegraph Company, 22, 69, 78
American Vacuum Society, 120
Ames Research Laboratory, 48, 88
Amtek, 88
Analog Devices, 92
Analog Devices Enterprises, 92
Anderson, John, 146
Anthony, John, 114, 121
Anthro Corporation, 45
Apollo Computer, 69, 71, 93
Apple Computer, 68, 93, 181
Arlington, Oregon, 176
Associated Oregon Industries, 134, 136, 143
Atiyeh, Victor, 82, 83, 113, 117, 118, 135, 138-43 *passim*, 151
Atlas Intercontinental Ballistic Missile Program, 17
Atteberry, Wayne, 170
AuCoin, Les, 133, 134
Audio Precision, 184

BancBoston Ventures, 92
Bave, Milt, 28, 29, 35
Bardeen, John, 7
Beaverton, Oregon, 6, 34, 47, 50, 66, 121, 122, 148, 151, 152, 169
Beaverton Area Chamber of Commerce, 173
Beaverton School District No. 48, 152
Beckman Instruments, 12, 14
Belleville, Logan, 35

Bell Telephone Laboratories, 7-8
Benedict, Donald, 103-105
Bipolar Integrated Technology, 91
Blake, Larry J., 119
Blakemore, John S., 108
Blumel, Jospeh C., 123
Boeing Company, 15, 20, 64, 65, 66
Bouton, Frank, 47
Brattain, Walter, 7
Broome, Oringdulph, 19
Brown, Frank, 12-13, 15
Brown Electro-measurements Company, 13
Brown Engineering Company, 12-13
Bruggere, Thomas, 67, 68, 70, 72, 119, 128, 134, 166, 182
Buehler, 63
Burroughs Corporation, 53, 88
Burrows, Mary, 144
Business-Education Compact of Washington County, 118, 152

CII Honeywell Bull, 53
California Automated Designs, Inc. (CADI), 68-69, 71-72
California Institute of Technology, 12, 14
Carlson, F. Paul, 108-109, 110, 114
Carter, Robert L., 47, 64
Cease, Jane, 153
Chao, Gene, 73, 114, 124
Charlesworth, Alan, 49, 50
Chevron, 49
Civil Rights Act of 1964, 178
Clean Air Act of 1963, 174
Coldwell Banker, 83
Colley, Stephen, 87-88
Communication Workers of America, 159, 160, 163, 164
Community Right to Know and Protection Act, 144
Computer Systems Architects, 86
Computer-vision, 72, 73, 74
Conkling, Gary, 121
Control Data Corporation, 48, 49, 54
Convex, 87, 190
Cornelius, Oregon, 18
Cornell Oaks Corporate Center, 170
Cornell University, 55, 133
Cossor, 36-37
Council for Advanced Science and Engineering Education/Research for Industry, 114, 117, 118, 120, 142
Cray Research, 85-90 *passim*
Crosspoint Venture Partners, 186
Culler, Glen, 48, 49

Daisy Systems, 67, 70, 192
Dallas, Warren "Dal," 35-36, 37
Data General Corporation, 47, 77
Daves, G. Doyle, 107
Davis, Bob, 35, 36, 37, 40-41
Davis, William E. "Bud," 115, 142
De Forest, Lee, 7
Diack, Samuel L., 102, 103, 108
Digital Equipment Corporation, 52-53, 93, 191
dilithium Press, 183
Dumont, 29, 31
Dyke, Walter P., 100, 108

ETA Systems, 87
Eastham, Melville, 12
Eberstadt, F., and Company, 23
Eckert, J. Presper, 8
Eisenhower, Dwight D., 99
Electro-Plasma, 76
Electro Scientific Industries (ESI), ix, 10, 13-16, 17-25, 46, 56, 62, 63, 102, 114, 121, 122, 123, 130, 146, 149, 162, 164, 177, 191
Elliott, Richard, 105
Elorriaga, John, 14
Encore Computer Corporation, 88
Enterprise Wallpaper Manufacturing Company, 51

Environmental Protection Agency, 131, 174-75
Epson Corporation, 80-81, 82, 158, 171
Equal Employment Opportunity Commission, 178
Eror, Nick, 105
Export Administration Act of 1979, 129
Export Trading Company Act of 1982, 130

Fadeley, Ed, 135, 136
Fairchild Camera and Instrument Corporation, 58
Fairchild Semiconductor, 8, 9, 58, 62, 175, 182
Far West Federal Bank, 184
Federal Aviation Administration, 65-66
Federal Environmental Pesticide Control Act, 174
Federal Express, 65, 66
Feuer, Michael, 69
First Interstate Capital, 92
Five Oaks Industrial Park, 170
Fletcher, Irv, 138
Flight Dynamics, 64-66, 177
Floating Point Systems, ix, 23, 43, 46, 47-55, 56, 64, 77-78, 86-87, 88, 89, 111, 122, 123, 133, 141, 173, 177, 179, 189, 190-91
Ford Microelectronics, 90, 191
Forest Grove, Oregon, 121
Frost, Chuck, 148
Fujitsu, 81-82, 84, 87, 90, 151, 158, 171, 191

GTE Company, 75
Gardner, Jim, 114, 138
General Dynamics (Convair), 17
General Electric, 7, 20, 21, 51, 54, 58, 76-77, 185
General Electric Medical Systems, 55
General Radio (GenRad), 12
Geosource, 49
Gigabit Logic, 90
Giles, Ken, 91-92
Gold, Michael, 177
Gray, John, 103
Green, Edith, 145
Greylock Management, 68, 186
Grid Systems, 76
Grove, Andrew, 58, 60, 63, 64, 167

Hambrecht and Quist, 23, 68, 93, 186
Hannman, Al, 37
Harris, 90
Hatfield, Mark o., 18, 19, 100, 101, 102, 104, 133, 145
Hauser, Frank, 106
Hawthorne Electronics, 35
Hawthorne Farm Industrial Park, 62
Hayes, Henry, 152
Hendrickson, Margie, 136
Hewlett, William, 7, 160
Hewlett-Packard, 7, 15, 16, 17, 42, 58, 69, 72, 83, 123, 160
Hill Family Foundation, 107
Hillsboro, Oregon, 6, 62, 63, 64, 73, 79-82 *passim*, 121, 148, 191
Hitachi, 40, 87
Hoff, Ted, 9, 59, 167
Holmes, Fred, 106
Hout, Eldon, 61
Hughes Aircraft, 90, 191
Hunt, James, 117
Hurd, James W., 74, 75

ITT, 90
Integrated Measurement Systems, 184
Intel, 56, 57, 58-64, 68, 77, 83, 87, 92, 93, 110, 114, 123, 133, 158, 160, 162, 165-66, 167, 181, 182, 191
Intergraph, 72

International Business Machines (IBM), 8, 49, 53, 55, 58, 68-71 *passim*, 75, 85, 87, 88, 141, 191
InterVan Partners, 92

Japanese Federation of Economic Organizations (Keidanren), 82-83
Jensen, James, 145
Jet Propulsion Laboratory, 88

KEX radio, 11
Karakas, Steve, 179
Karakas, VanSickle and Ouellette, 179
Katz, Vera, 113-14, 118, 140, 141, 142
Keller, Ira, 103, 105-107
Kerans, Grattan, 138
Kerr, Richard, 106, 107
Kilby, Jack, 8, 58
King, Chin, 75
Kinship, Earl, Capital, 186
Kitzhaber, John, 140, 141
Kleiner, Perkins, Caufield and Byers, 93
Kobbe, John, 30, 31, 35
Kumakura, Takashi, 40
Kyocera Corporation, 82

LaFalce, John J., 132
Lamoreaux and Glynn, 68
Land Conservation and Development Commission (LCDC), 149, 152, 171
Langeler, Gerard, 67, 70
Lattice International, 77-78, 189, 191
Lawson, Harold, 11
Lee, Marshall, 29, 33
Lewis and Clark College, 105
Lidke, John, 152
Lintner, Paul D., 22, 114, 121
Lintner, Paul D., Center for Advanced Education, 121, 122
Lockheed-Georgia Company, 66
Lohja, 76
Louisiana-Pacific Corporation, 77-78

McCall, Tom, 56, 146
McCormick, Pat, 134, 135, 141
McCracken, Mike, 144
McDowell, Glenn, 26, 28, 29, 35
McGregor, Douglas, 17
McHenry, Julie, 179
Massey, Gail, 106, 107
Matsushita, 75, 76
Mauchly, John W., 8
Mayhew, Larry, 44
Mentor Graphics, 66-72, 73, 93, 119, 123, 128, 164, 168, 177-82 *passim*, 192
Merlo, Harry, 77
Metheus-CV, 73-74, 180
Metheus Corporation, 72, 73-74, 114, 124, 189-90
Metoramic Sciences, 22-23
Metropolitan Service District, 149, 151, 153, 154
Mikesell, Howard, 40
Mikros, 19-20, 22
Millikan, Robert A., 12
Mitsubichi, 90
Moffenbeier, David, 67, 70
Mondale, Walter, 132
Moore, Gordon, 57, 58, 60, 62, 64, 160, 167, 182
Morgan, Merle, 14
Morita, Akio, 40
Morrow Electronics, 66
Motorola, 91, 191
Moulton, Cliff, 35
Moyle, Kenneth, 114
Murdock, Jack, ix, 10, 16, 26-28, 29, 35, 37, 38, 40-41, 43, 160, 161, 169, 172, 173
Murdock, M. J., Charitable Trust, 107-108, 173

Murdock Radio and Appliance, 26-27

NCube, 87-88
Nan, Ning, 68-69, 71
National Aeronautics and Space Administration (NASA), 15, 48, 50, 65, 88, 145-46
National Cooperative Research Act of 1984, 131
National Defense Education Act of 1958, 99
National Environmental Policy Act, 174
National Science Foundation, 86, 107, 178
Nippon Electric Company (NEC), 79-80, 82, 83, 87, 147, 158, 171
Northrop Research and Technology Center, 87
Northwest Air Pollution Laboratory, 107
Northwest Electronics Workshop, 24
Northwest Instruments Systems, 180
Northwest Science Investment Corporation, 102
Northwest Venture Capital, 186
Noyce, Robert, 8-9, 57, 58, 60, 62, 64, 133, 160, 166-67, 182, 191

Oak Ridge National Laboratory, 88
O'Leary, George, 49, 54
Omark Industries, 103, 108
Oregon Center for Advanced Technology Education, 118, 119-20, 122, 123, 142, 156
Oregon Consortium for High Technology Education, 113, 117, 118, 142, 146
Oregon Economic Development Department, 83-84, 146-48
Oregon Educational Coordinating Commission, 112, 114, 116
Oregon Graduate Center, 18, 100-12, 115, 117, 120, 123, 133, 156, 173, 174, 177,178, 192
Oregon Graduate Center Corporation, 109, 110
Oregon Graduate Center Science Park, 108-10, 170
Oregon Institute of Technology, 119
Oregon Medical Research Foundation, 102
Oregon Museum of Science and Industry, 121, 174
Oregon Public Employees Union, 135
Oregon Regional Primate Center, 100-101, 173
Oregon State Board of Higher Education, 101, 102, 113
Oregon State Department of Environmental Quality, 143-45, 151, 174, 176-77
Oregon State Department of Planning and Development, 100
Oregon State Resource and Technology Development Corporation, 186
Oregon State System of Higher Education, 105, 113-20 *passim*, 141-42
Oregon State University, 100, 102, 113, 116, 120, 122-23, 145, 156
Oregon Student Public Interest Research Group, 143
Ouellette, Pierre, 179

PM Industries, 123
Pacific Hybrid Microelectronics, 123
Pacific Northwest Bell, 141
PacifiCorp Ventures, 186
Pacific Power and Light, 65
Pacific Telecom, 65, 66
Packard, David, 7, 160
Palmer, John, 87, 88
Palomar Systems and Machines, Inc., 22-23

Pankow, James, 177
Patz, Alan D., 185
Paulson Investment Company, 186
Pennington, Derrol, 35
Pentrix, 41-42
Perkins, Coie, Stone, Olsen and Williams, 179, 180
Perry, Louis, 114
Phosphon, 76
Photonics, 76
Planar Systems, 45, 74-77, 109-10, 151, 180, 186
Plessey Research, 90
Polits, Bill, 35
Portland Chamber of Commerce, 136
Portland City Club, 102, 187
Portland Community College, 113, 114, 120, 121, 123
Portland Development Commission, 106, 154
Portland General Electric Company, 145, 186
Portland State University, 99-100, 101, 113, 115, 116, 118, 120, 123, 142, 156
Potash, Hanam, 89
Powell, Casey, 92, 182
Prince, Tom, 47

Quadrant, 170
Questech, 65

Radio Corporation of America (RCA), 7, 29, 31, 76, 186
Rainier Venture Partners, 186
Rasmussen, Linda, 164
Raytheon Company, 92, 185
Reagan, Ronald, 127, 137
Reed College, 99, 100
Reedville, Oregon, 151
Regis McKenna, 63, 179
Rembold Corporation, 108, 109, 110
Resource Conservation and Recovery Act, 175
Rhiger, Dick, 35
Richardson, William, 87, 88
Riviera Motors, 170
Robinson, Philip R., 114
Rock, Arthur, 58
Rockwell, 75, 90
Rockwood, Lawrence B., 14
Roe, David, 105
Roehr, Frank, 179
Ropiequet, Dick, 35
Rothschild, L. F., Unterberg and Tombin, 68

Salquist, Jon, 54
Samsung, 191
San Diego Supercomputer Center, 89
Schumann, Robert, 88, 89

Scientific Computer Systems, 88-90
Sea-Port Industrial Group, 170
Securities and Exchange Commission, 23
Seiko Epson Corporation, 80
Sequent Computer Systems, 88, 92-94, 170, 179-86 *passim*, 192
Sharp, 75, 76
Shaw, Ralph, 168, 183, 187
Shaw Venture Partners, 93, 183, 186
Shell Development Company, 88
Shimizu Construction Company, 82
Shockley, William, 7, 8, 57-58
Shockley Semiconductor Laboratory, 8, 57, 62
Short, Robert, 145
Siemens AG, 22, 76, 94
Sigmatron Nove, 76
Silicon Valley, 60, 131, 150, 152, 156, 166, 167, 175, 179
Small Business Innovation Development Act, 126
Sony Corporation, 40
Sony-Tektronix Corporation, 40

Sparcatron, 69
Sperry Rand, 50, 53
Standard Insurance Company, 81, 114, 170
Stanford Research Institute, 103
Stanford University, 58, 71, 88
Stanford University Science Park, 18, 108
Star Technologies, 54, 183
State Farm Insurance Company, 23
Stoel, Rives, Boley, Fraser and Wyse, 179
Strain, Clayton, 13, 14, 18
Strain, Douglas, ix, 10, 11-19 *passim*, 24, 25, 46, 102, 146, 164
Strain, Leila (Mrs. Douglas), 12
Straub, Bob, 62
Sud, Rahul, 77, 78
Sullivan, Richard H., 100
Sunset Corridor Association, 56, 153-54, 169-71, 172
Sunset Scientific Park, 18-19
Sutter, Hill, 68, 186
Suwa Seikosha Company, 80
Swerling, Stephen, 69
Synatron Corporation, 82

TA Associates, 89
Tanasbourne Commerce Genter, 81
Tarr, Robert H., 102
Tax Reform Act of 1984, 127
Teeples Electronics, 14
Tekintag, 39
Tektronix, ix, 7, 10, 15-18 *passim*, 23, 24, 26-45, 46, 53, 56, 63, 72, 75, 103, 104, 105, 110, 114, 118, 121, 123, 128-29, 130, 134, 137, 138, 158, 177-78, 179, 189, 191-92:
acquisitions, 41-42; field offices, 36; labor relations, 160-69 *passim*; move to Washington County, 34, 172; overseas operations, 36-38, 39-40, 45; philanthropy, 172-73; relations with local government, 148-49, 152, 154, 175-76; spin-off companies, 43, 45, 47, 67, 74, 90-91, 182, 184, 185-86
Tektronix Development Corporation, 91, 185
Teradyne, 93
Terman, Frederick, 18
Texas A & M University, 122
Texas Instruments, 8, 56, 90, 191
Thompson, Ross, 114
Thurow, Lester, 132
Tigard, Oregon, 49, 82, 121
Tippery, Miles, 26, 27-28, 29, 35
Tomita, Masakatsu, 81
Toshiba, 40
Towne, James, 73
Tri-County Metropolitan Transportation District of Oregon (Tri-Met), 149, 154
TriQuint Semiconductor, 45, 90-91, 185-86
Tualatin, Oregon, 121
Tualatin, Valley, 3-6, 9, 146, 151, 154
Tualatin Valley Economic Development Corporation, 56, 171
Turner, Lloyd D., 53-54, 55, 190

Umpqua Community College, 115
Union Venture Corporation, 92
U.S. Air Force, 17-18, 19, 107
U.S. Army, 76, 107
U.S. Bancorp First Source Capital, 186
U.S. Bureau of Census, 8
U.S. Department of Commerce, 129, 130
U.S. Department of Defense, 14, 17, 86, 90, 129
U.S. Forest Service Radio Development Laboratory, 10-11, 12, 25
University of California at Berkeley, 88

University of Michigan, 88
University of Oregon, 100, 102, 113, 116, 118, 120, 122-23
University of Portland, 99

Valid Logic Systems, 67, 70
Vancouver, Wash., 82
VanSickle, Sharon, 179
Varian Associates, 20, 29
Venrock Associates, 68, 93
Vollum, Howard, ix, 10, 11, 16, 26-32 *passim*, 35, 37, 38, 40-45 *passim*, 75, 100, 103, 104, 108, 160, 161, 169, 172, 174
Vollum, Jean (Mrs. Howard), 174
von Neumann, John, 8

Wachovia Bank and Trust, 170
Waggener, Melissa, 179
The Waggener Group, 179
Walker, William, 75
Wallen, John, 38
Wantland, Earl, 37, 39, 43, 118, 192
Washington County, ix, 6, 18-19, 56, 60-64, 81, 83-84, 104, 108, 110, 119-20, 121-22, 148-54, 155, 156-57, 169-77, 184-85, 192-93
Washington County Board of Commissioners, 150, 151
Water Quality Act of 1965, 174
Western Electric, 79-80
Western Geophysical, 49
Western Kraft, *see* Willamette Industries
Westinghouse, 20, 90
West Union Park, 170
Weyerhaeuser Corporation, 106, 107, 170
Willamette Industries, 103, 106
Willner, Don S., 101
Wilson, George, 91
Wilson, Kenneth G., 85, 86
Wilsonville, Oregon, 44, 88
Winningstad, C. Norman, ix, 42, 43, 46-49, 53, 54, 55, 78, 141, 173, 190-91
Winningstad, Dolores (Mrs. C. Norman), 173-74
Woods, William, 106
World Affairs Council, Portland, 138
Wyden, Ron, 137-38

Yale University, 88
Young, Jeremy H., 179
Young and Roehr, 179

Zschau, Edward, 126

Colophon

The typeface used for both text and display in *The Silicon Forest* is Sabon. Sabon was created in the early 1960s by Jan Tschichold, one of the twentieth century's foremost type designers. Typesetting was done by The Typeworks of Point Roberts, Washington. Printing and binding of *The Silicon Forest* was performed by Edwards Brothers, Inc., Ann Arbor, Michigan. The book is printed on 50 lb. Sebago paper, an archival and recycled stock.

This book was designed and produced by the Oregon Historical Society Press.